The Perceptory Force

A Structural Theory of Consciousness and Reality

In endless possibilities,
we all are that we perceive.

Joanna Accordi

The Perceptory Force

A Structural Theory of Consciousness and Reality

ISBN: 9798181530638

First Edition, 2026

Independently published

Dedications

In the memory of all who sacrifice,

I dedicate all that I create.

To My Beloved Children

Alice & Jackson

From Perceptory Networks You were Born

The Perceptory Network You Have Become

Always Be

You are so Perfect to Me

Rocket Your Thoughts on the Fly

Never let them cut Your Sky

Love and Live Unconditional Infinity

A Paradox of Your Reality

Where You Belong

Always Yours

Mum

To You

As you read this sentence you are subconsciously aware that this text is written by somebody. Based on all the information that you have gained at this very point, the concept of me, the writer, is now resurfacing in your conscious mind. Now you smile because you know I'm right. Your facial muscles confirmed my existence.

You could interpret this situation as me bypassing the non-existent time and space, and directly manipulating your facial muscles by this very sentence that I am now typing.

So where is the time and where is the space that divides us?

Enjoy Entangled Human Being

Contents

Introduction

The more we evolve and the more we perceive, the harder it becomes to apply mathematical representations to the conscious phenomena we operate within and via. It's not that theoretical physicists and mathematicians have avoided or given up on the pursuit of consciousness; rather, they haven't identified the Perceptory Force as a fundamental component of anything that exists, nor as a fundamental component of magnetism and electromagnetism. As we continue to explore and make new discoveries, we find that previously accepted theories and equations must be continuously replaced with new ones that better fit reality.

We could even state that numbers themselves, which serve as a means of representing patterns and relationships, possess inherent limitations in fully encapsulating the entirety of existence. Mathematics evolved because humans needed to predict uncertain outcomes. It is a logical language built on the foundation of probabilistic prediction, enabling us to interpret and navigate the world's intricacies. But prediction is not perception, and a framework built to anticipate outcomes will never fully describe the force that brings those outcomes into existence in the first place.

The continuous process of discovery through new approaches is crucial for the advancement of our consciousness, even when logic and abstraction are not enough in the pursuit of the unknown. We are limited by our own thoughts.

Refining our understanding of consciousness itself, through questioning it, is an essential part of the philosophical and scientific process. This, in turn, propels all of us forward and opens the gates to a completely new way of thinking and perceiving. "The 10 Laws of Consciousness" represents a new theory and hypothesis that provides a fresh framework for understanding the universe.

A reader who comes to this book from physics, philosophy of mind, cognitive science, or cosmology deserves to know at the outset what kind of book it is.

This is a structural account. Its claims describe how things are arranged and how they relate, not how much of anything there is, and so they are stated in prose rather than in equations. Where the framework does reach toward mathematics, it does so symbolically, and those symbols are gathered in Appendix B. They are a beginning, an invitation to the formal work that could follow, not a finished calculus.

The book does not set itself against quantum field theory, general relativity, or the Standard Model. Those theories describe their domains with a precision this book does not attempt to match or replace. What it offers instead is a single architecture beneath them: a way of seeing why they take the shape they do, how they connect to one another, and how they connect to consciousness itself. The work of turning that architecture into numbers is still to be done; this book maps the territory that work would formalize.

Even so, the framework does not hide behind generality. It commits to specific things that can be looked for, and whose absence would count against it: a particle I call the A-particle, the restoration of parity at high energy, a three-fold regularity in the structure of the cosmic web, and others. These are set out in Appendix C.

One word needs care before we begin. Throughout this book, *consciousness* names a structure, the closed loop in which a *Perceptory Field* perceives itself. This is broader than the sense the word usually carries in the study of mind, where it points to subjective experience, to what something is like from the inside. What molecular, cellular, organismal, or cultural consciousness might feel like from within is a question I leave open. When those phrases appear in these pages, it is the structure I mean.

Part I · Foundations

What is Consciousness, Perception and Reality?

Consciousness constitutes the most intimate fact accessible to any reader of this book, and simultaneously the least understood. Every reader possesses it; no reader can define it. We enter it upon waking without effort, relinquish it in dreamless sleep without noticing, and inhabit it throughout a lifetime without the capacity to step outside it and examine its structure from without. It is the singular datum that cannot be doubted, the act of perceiving is itself undeniable, and simultaneously the phenomenon that resists every attempt at derivation from more fundamental principles.

Before surveying what others have established, I will articulate what I have been compelled to conclude from my direct experiences and observations gathered at the boundaries of consciousness, where its properties become more evident than they are in ordinary conditions.

This asks me to do something a work of theory rarely does: to begin with my own life. I am aware of the objection — that one person's experience is no foundation for a claim about the structure of consciousness itself. But a structure tends to be invisible while it holds and becomes legible only when it is strained, fractured, or suspended; the experiences that showed me consciousness most clearly were the ones in which my own came apart and had to reassemble. Some of what follows is extreme, and some of it is hard to set down. I record it not as confession but as evidence — as the report of an observer noting the conditions under which a phenomenon became visible.

Formation

My own consciousness did not begin at any identifiable moment, for it precedes the formation of episodic memory itself. Awareness and recognition of my mother's voice occurred prior to birth, which indicates that the apparatus of perception was already operational in utero, prior to the full development of my cortical structures to which mainstream neuroscience typically attributes conscious experience. My earliest retained memory is of biting my own feet in infancy, and the persistence of that memory constitutes evidence of self-reference preceding verbal cognition. Consciousness, in other words, was not contingent on the maturation of the brain; it was already present, and the biological architecture developed around an awareness that preceded it.

Psychological Trauma

Consciousness has also revealed itself to me at its structural limits. Following the abduction of my children, and through the period of sustained abuse that preceded that event, I experienced what can be designated identity collapse, but the collapse was not a termination of consciousness. It was a fragmentation of what had been, before the trauma, a single tightly integrated Perceptory Network into distinct fields, each with its own characteristic operating mode: a child, a teenager who held the humor, a mother, and others. Before the fragmentation, the integration had been so complete that I did not perceive it as integration; the prior unity became visible only by its loss.

The six fields were co-present, entangled within a Perceptory Network, and each one was capable of independent operation. In ordinary daily life, one field operated at a time, typically the most energy-efficient mode for the situation at hand. Under conditions I could regulate, two fields could communicate with each other: the child speaking to the mother, or two of them consulting when a hard decision required input from more than one perspective. Each one registered the others as other. They were not voices within a single self; they were distinct fields entangled within the network that held them. I understood thought as the byproduct of the communication between two perceptory fields, born on the boundary between the two.

What I could also do, and what made reintegration possible, was observe all six simultaneously while moving between them. That observing perspective was not a seventh field above the six. It was the network itself perceiving its own constituents, the higher-scale Perceptory Field whose content is other Perceptory Fields, looking at its own structure from within. Before the trauma, the network-level perspective and the field-level operation had coincided so closely that I did not experience them as distinct. The fragmentation pulled them apart, and in doing so made the network's architecture visible to itself.

The dissociation itself, I now read as a Perceptory Network event rather than a single-field failure. When input exceeds what the network can integrate, the network can sever or freeze the entanglements among its constituent fields, sealing each field off, halting cross-field registration, refusing further data intake. It is not a breakdown but a protection. The network buys time to reorganize by temporarily refusing its own coherence.

Reintegration, in my case, was not a return to the pre-trauma configuration. It was the re-establishment of tight coupling among the six fields with one of them, the calmest, the most energy-efficient, becoming the dominant operating node through which the network now typically functions. The child-field was reabsorbed so fully that her separate operation is no longer accessible to me. The others remain available as modes I can deliberately enter when the situation calls for them. The post-reintegration network is not the pre-trauma network restored; it is a new architecture that includes what the fragmentation taught the network about itself.

Concussion

I have also traversed the complete suspension of consciousness through concussion, following a motorbike accident. The recovery from that suspension was not gradual but stratified. For a prolonged interval afterward, perceptual intensity was markedly amplified across every sense at once: chromatic saturation deepened, auditory texture sharpened, object boundaries took on an almost crystalline definition, smell became unusually acute, and visual resolution improved beyond its prior baseline. The instrument of perception had been disrupted and, in its restoration, briefly made itself visible, as though the ordinary filtering mechanisms had not yet fully re-engaged.

What I did during that window matters for what came of it. Some weeks after the accident, with the heightened state still in place, I began sustained high-intensity intellectual work, deliberately loading the network with demanding cognitive tasks while its plasticity was still elevated. The network reorganized around that load. When the window eventually closed and ordinary filtering returned, what remained was not the pre-concussion network restored to baseline but a network with different access to itself. Memory in particular became directed: I could move recall toward early childhood, or toward any period I chose, with a fluency the pre-concussion network had not possessed. The gain held.

I read this, retrospectively, as a Perceptory Network rebuilding itself under conditions where its entanglements were loose enough to be reshaped before they re-solidified. The concussion opened a structural window; the intellectual work was the load that determined what the network became when that window closed.

The reconstruction was not deterministic. It could be steered, and what steered it was the network's own activity within the plasticity phase. There was no separate agent doing the steering. The network shaped itself by what it did while it was reshaping.

General Anaesthesia

General anaesthesia constitutes a further and distinct phenomenon. Under it, no memory trace is laid down; the interval between induction and emergence is not experienced as duration but is simply absent from the record. Consciousness does not slow or thin during that interval; it does not register at all. The substrate that ordinarily produces continuous self-perception is, for a measured pharmacological window, decoupled from itself.

A related set of observations concerns molecular hydrogen and its effect on the perceptual network. Ingested prior to dental procedures, H_2 appears in my experience to interfere with the efficacy of local anaesthetics, the numbing onset is delayed, the depth of anaesthesia diminished, and nociceptive signals continue to register where they would ordinarily be silenced. Ingested in ordinary conditions, with no anaesthetic present, H_2 produces a global shift in perceptual state: a noticeable calming, and a heightening of perceptual acuity that I can register from inside. The two observations, taken together, point toward a single biochemical action operating at two scales.

The pharmacological literature provides the mechanistic context. Molecular hydrogen is a selective antioxidant that scavenges hydroxyl radicals and peroxynitrite, the two most cytotoxic reactive species in cellular biochemistry, while leaving other reactive species, such as hydrogen peroxide and nitric oxide, untouched. Local anaesthetics interact with the same class of reactive species, with peroxynitrite as a shared targ et. When H_2 is present in tissue prior to anaesthetic administration, the reactive-species pool with which the anaesthetic would ordinarily engage has already been partially scavenged, and the local biochemical environment in which the anaesthetic operates is altered. At the systemic scale, H_2 crosses every biological membrane rapidly, including the blood-brain barrier, and the same scavenging action proceeds wherever the targeted species are produced, including continuously in the brain, where mitochondrial activity generates them as routine metabolic byproducts. The calmer, more perceptive state I notice after ingesting H_2 is consistent with a reduction in the cortical oxidative burden that ordinarily produces a measure of perceptual noise and excitability.

Whether these are the mechanisms of the effects I have observed is not established, and the question deserves clinical investigation. What the observations make clear, regardless of mechanism, is that whatever consciousness is, and whatever permits nociceptive and perceptual signals to register within it, is coupled to biochemical conditions in ways that can be deliberately modulated, and that the network's perceptual properties at local and systemic scales are coupled to the same underlying chemistry.

These are not arguments in the formal sense. They are observations compiled from a life that has been compelled to encounter consciousness across a wider range of states than most individuals' traverse. They have brought me to a position that no textbook could have produced: consciousness is not generated by the brain in the sense of matter producing mind. Consciousness is generated structurally wherever perception closes upon itself; the brain is one biological substrate through which such closure expresses under certain conditions. Disruption of the apparatus modifies the expression, but consciousness itself is not constituted by neurons, and no exhaustive mapping of neural activity will locate its origin.

Others' philosophical and scientific frameworks

This is not the conclusion most philosophical and scientific frameworks have reached, though certain of them have approached its periphery.

The Western philosophical tradition of consciousness originates, for present purposes, with René Descartes. In 1641, in the course of stripping away every belief susceptible to doubt, Descartes arrived at a single irreducible proposition: cogito, ergo sum, the act of thinking guarantees the existence of the one who thinks. Even under the hypothesis of universal illusion, the act of perception is itself undeniable. Descartes took this as demonstration that mind and body constitute distinct substances, a dualism that has structured Western thought ever since.

The difficulty is that Cartesian dualism generates an interaction problem that no subsequent philosopher has resolved. If mind and matter are genuinely separate in kind, the mechanism by which they influence one another becomes inexplicable. How is a thought converted into the movement of a hand? How is the incidence of light on the retina transformed into the subjective experience of colour?

The contemporary formulation of this difficulty was articulated by David Chalmers in 1995, who designated it the hard problem of consciousness. The so-called easy problems, in Chalmers' analysis, are empirically tractable: the mechanisms of stimulus discrimination, information integration, and verbal reporting are amenable to neuroscientific investigation. The hard problem is categorically distinct. Why should any of this neural activity be accompanied by subjective experience whatsoever? Why is there something it is like to be a conscious organism, rather than merely information processing in the dark? A functionally equivalent brain, executing identical computations, could in principle operate without any accompanying inner life. Yet the human brain does not. The question of why remains unresolved.

Three decades after its articulation, the hard problem persists, not for want of sustained philosophical effort, but because the framework within which the effort is conducted may itself constitute the obstacle. Every attempt to derive subjective experience from objective matter has encountered the same impasse. Certain philosophers, notably the phenomenologists Husserl and Merleau-Ponty, have pursued an alternative trajectory. They have maintained that consciousness is never a property in isolation; it is always consciousness of something. Perception, in this account, is not an event occurring within a self-contained mind, but the very structure by which a world becomes manifest at all. This formulation approaches the framework developed in this book, and merits acknowledgment. Yet phenomenology stops at description. It articulates the structure of experience with precision but does not pose the further question: whether perception itself constitutes a fundamental force operative within the universe.

Where philosophical inquiry has pursued the question of why, neuroscience has pursued the question of how. Over the last half-century, empirical research has sought what are termed the Neural Correlates of Consciousness, the specific patterns of brain activity that reliably accompany conscious experience. This programme has produced substantive findings: the thalamus and specific cortical networks are demonstrably essential; particular regimes of neural synchrony correlate with awareness; lesions in designated regions abolish specific aspects of conscious experience.

From this empirical foundation, several theoretical frameworks have emerged, each rigorous, each partial.

Integrated Information Theory, developed by Giulio Tononi, proposes that consciousness is mathematically identifiable with integrated information, specifically, the quantity of information a system generates that exceeds the sum of its constituent parts. Tononi formalizes this as Φ (phi), a measurable quantity. Systems with high Φ are correspondingly high in consciousness; systems with Φ of zero possess no experience whatsoever. The theory is mathematically elegant and, in principle, empirically testable. It does not, however, explain why integrated information should be accompanied by subjective experience rather than merely constituting unconscious processing.

Global Workspace Theory, proposed by Bernard Baars and subsequently developed by Stanislas Dehaene, characterizes consciousness as a broadcast phenomenon. The majority of cortical processes operate locally and without awareness; when information is distributed across a brain-wide network, rendered globally available to attention, memory, language, and executive function, it enters consciousness. This framework accounts for a considerable range of empirical observations about the dynamics of conscious access. It does not, however, explain why the broadcast should be accompanied by felt experience.

Predictive Processing, associated with Karl Friston and the free energy principle, construes the brain as a generative prediction apparatus. The brain continuously generates models of anticipated sensory input, compares these predictions against incoming data, and updates itself to minimize predictive error. Consciousness, on this account, is the ongoing construction of a model of world and self. This framework is powerful, and notable for its structural resemblance to the feedback architecture this book will develop in the later Laws. Yet Friston himself acknowledges that predictive processing describes the dynamics of cognition without establishing why those dynamics are accompanied by experience.

Each of these frameworks, however sophisticated, ultimately describes correlations. None accounts for why there is an interior dimension to any of it. This is not incidental. Neuroscience proceeds from the foundational premise that consciousness is produced by the brain. If that premise is mistaken, if consciousness is not generated by matter but is more fundamental, with matter constituting one of its manifestations, then no degree of neural mapping, however comprehensive, will identify the source. The cause is not locatable in the effect.

Physics occupies a different position in relation to the problem. For most of its history, physics excluded consciousness from its domain entirely. The observer was modelled as a neutral recorder, a mind positioned outside the equations, extracting measurements from instruments. Physics described the world; consciousness was consigned to other disciplines.

Quantum mechanics disrupted this arrangement. The double-slit experiment and the subsequent family of related experiments revealed a phenomenon for which classical physics had provided no preparation: the act of observation alters the observed. A particle's behaviour is contingent on whether measurement is being performed. The Observer Effect is now a foundational feature of quantum theory, however philosophically uncomfortable it remains. A minority of physicists accept it at face value and investigate its implications for the role of consciousness. The majority have sought mechanisms by which to explain it away, decoherence, many-worlds branching, hidden variables. The philosophical discomfort is legible in the sheer proliferation of proposed alternatives.

A small number of physicists have been more willing to engage the implications directly. John Archibald Wheeler articulated the concept of a participatory universe: a cosmos in which observers are not external to reality but constitutive of it. His formulation "it from bit" proposed that the physical world arises from information, and that information presupposes an observer to register it. Wheeler's thinking gestured toward the possibility that perception is foundational, though he did not formalize it as a force. Roger Penrose, in collaboration with the anaesthesiologist Stuart Hameroff, advanced the hypothesis of Orchestrated Objective Reduction. Their proposal was that consciousness arises from quantum processes within neuronal microtubules. The theory is controversial and has been subject to substantial criticism. What it articulates correctly, and what is pertinent here, is the intuition that consciousness is not reducible to classical neural computation. Something more fundamental is operative, something that engages the quantum stratum of reality itself.

Physics, in short, has been compelled by its own findings to recognize that the observer is implicated in what is observed. It has not proceeded to take the further step: the question of whether perception itself constitutes the foundation of what appears as observable.

Philosophy inquires into the emergence of subjective experience from objective matter and finds no demonstrable answer. Neuroscience correlates consciousness with neural activity and cannot establish why the activity is accompanied by experience. Physics concedes that the observer is implicated and cannot specify the nature of that implication.

The impasse exhibits a consistent structure. Every framework attempts to derive consciousness from something more fundamental, brain, information, matter, computation, and every framework encounters failure at the same juncture. The question that has not yet been posed is whether the direction of derivation has been inverted from the outset.

What if consciousness is not produced by matter, but matter is perceived into existence?

If that is the case, what is it that performs the perceiving?

To address that question requires reconfiguring the semantic field of a term long constrained by narrower usage. In ordinary discourse, perception designates the operation of the sensory modalities: vision, audition, tactile discrimination, olfaction, gustation. It is associated with sensory organs and with the cortical processing of sensory input. This is not the sense in which I intend the term. Sensory perception is one manifestation of perception, the form most familiar to human beings, but it is not perception itself.

Perception, in the foundational sense developed in this book, is phylogenetically and ontologically prior to sensory apparatus. It precedes the evolution of nervous systems. It is more fundamental than the biological architecture through which organisms register their environments. Perception, properly understood, is the act by which anything that exists is registered as existing. It is the relational event at which presence is distinguished from absence. Wherever a relation obtains between two systems, any relation, of any kind, perception is operative.

A hydrogen atom encountering an oxygen atom and forming a covalent bond has, in this formulation, perceived it. A protein reconfiguring its conformation in response to the biochemical environment has perceived that environment. An immune cell recognizing a pathogenic signature has perceived the pathogen. A maternal system registering and integrating the cellular signals of the foetus has perceived those signals. None of these events requires visual, auditory, or neural apparatus. Each requires only that one system register another, and adjust itself accordingly.

This is the force I am asking the reader to apprehend: not the sensory faculty that evolved in biological organisms, but the more fundamental act of which that faculty constitutes a late and elaborate expression. In its formal designation within this book, this force is named the Perceptory Force. For the moment, I am asking only that the term perception be permitted to expand beyond its sensory connotation, until it encompasses every act of registration obtaining between any two entities that exist.

Now the analytical move that this entire chapter has been constructed to enable:

If perception is the fundamental act, then consciousness is what is structurally generated when perception closes upon itself. "Emergence" in the sense used here is not derivation from a prior material substrate but structural appearance: consciousness is what closure looks like from inside the loop, not a property produced by something more fundamental than itself. When a system perceives not merely its environment but perceives its own perceiving, when the act of registration registers itself, a self-referential loop is established. Within that closed loop, awareness is generated. The loop is what constitutes the interiority.

This is why the philosophical and scientific frameworks surveyed above have encountered their respective impasses. They have attempted to derive consciousness from matter, from neural activity, or from information processing, operating under the presupposition that perception is simply what brains do. The present framework inverts this presupposition. Perception is not a property of brains. Brains are the form perception assumes when it closes upon itself within biological substrate. Consciousness is not produced by the brain. Consciousness is perception perceiving itself, expressing through whatever apparatus is available.

This is why an infant can be conscious prior to the acquisition of language. This is why consciousness persists through identity collapse, through dissociation, through the disruption of the continuous narrative self: the loop is structurally more fundamental than the content instantiated within it. This is why anaesthesia interrupts consciousness without destroying it, the chemistry disrupts the loop without affecting whatever the loop is constituted from. This is why the observer in a quantum experiment cannot be separated from the measurement. The observer is perception itself, and the particle is what perception has produced.

Consciousness emerges from perception. Perception does not emerge from anything prior. Perception is the ground.

Which brings the argument, finally, to reality itself.

If perception is foundational and consciousness emerges from it, then reality cannot be what the prevailing frameworks have represented it to be. Reality is not a stage upon which consciousness is positioned as observer. Reality is not an independently existing world of objects subsequently registered by observers. Reality is the field of all perceptory possibilities, the totality of what can be perceived, what is being perceived, and what is performing the perceiving.

Reality is not what physics has told us it is. Reality is the dance between Nothingness and Infinity, held together by perception.

Every question concerning origins, how did the universe begin, from what did it arise, what preceded it, presupposes that reality is a thing that commenced. This presupposition constrains cognition into the assumption that everything must have a beginning and, therefore, an end. Reality, however, has neither beginning nor ending, because reality is not a thing. Reality is the continuous act of perception that differentiates what exists from what does not.

Two concepts must be oriented before the analysis can proceed.

Nothingness is not empty space, nor the void antecedent to creation. Nothingness is the absence of anything whatsoever, no matter, no energy, no dimension. It is infinite but it is not Infinity. It cannot interact; it cannot perceive. Yet for Nothingness to exist at all, it must be perceived, because the absence of something is recognizable only through the perception of what is present. This paradox constitutes the foundation of everything that follows, and I will develop it in full in the First Law of Consciousness.

Infinity is not endlessness in the conventional sense, an unbounded series of numbers, or a universe expanding outward without limit. Infinity is all possible perceptory distributions, perceiving Nothingness. It has no magnitude. It has no duration. Infinity simply is. I will develop this in full in Law X, where the Two Infinities, perceptual and propagation, are derived.

Between these two, the unperceivable ground of Nothingness and the total simultaneity of Infinity, lies everything we encounter as real. Reality is the dance between them, and perception is what enables the dance to occur.

From this framework, the Big Bang should not be interpreted as an absolute creation event from Nothingness, but as an observable expansion phase within an already perceptory reality. The designation "Big Bang" itself suggests an explosion localized at a specific time and place, with a centre, but reality has no centre. Networked galaxies diverge from one another without a central locus; the phenomenon of expansion manifests everywhere simultaneously. The framework does not deny the observational evidence of cosmic expansion; it disputes the metaphysical reading of that evidence as an absolute beginning. Nothing began. Possibilities are being perceived. The Big Bang names a phase of expansion within an infinite perceptory architecture, not the moment of creation from Nothingness.

The universe is not a thing that began. It is an infinite number of perceptory possibilities expanding within Infinity, where the perception of Nothingness serves as the fundamental building block of all that exists.

This is the framework the Ten Laws will now construct, proposition by proposition. The Laws do not contradict what philosophy, neuroscience, and physics have established. The Observer Effect is real, the Neural Correlates are real, the hard problem is real. The Laws reorder these findings. They propose that each of these fields has been describing manifestations of a more fundamental reality throughout, and that what appears impossible to derive from matter becomes evident once matter itself is understood as a perceptory phenomenon.

Part II

The Ten Laws

I Law of Consciousness

For Nothingness to exist, it must be perceived.

Orientation

Law I is the foundational Law of the framework. Every Law that follows depends on what is established here. The Law states a paradox so structurally tight that, once accepted, it forces the rest of the architecture into existence: for Nothingness to exist, it must be perceived. The remainder of the book unfolds from this single proposition. Law I therefore must do more than state the proposition; it must derive it, defend it, locate it within the philosophical traditions that have approached the question of Nothingness, and reveal the structural mechanism by which the perception of Nothingness produces the first act of self-perception that grounds everything else.'

The problem of Nothingness in the Western tradition

The framework's claim about Nothingness does not arise in a philosophical vacuum. Western philosophy has approached Nothingness from several directions, each of which the framework can position itself against in order to clarify its own contribution.

Heidegger, in Was ist Metaphysik? and elsewhere, identified Nothingness as the structural other against which Being defines itself. His question, why is there something rather than nothing?, is the question Law I answers, but Heidegger's framework leaves Nothingness as the unthinkable ground of metaphysical anxiety, the experience of which reveals to the existing being its own existence. The framework converges with Heidegger in recognizing Nothingness as foundational to the question of existence, and diverges in identifying Nothingness as a specific structural object whose existence depends on a specific structural act, perception, rather than as the existential horizon of finite being.

Sartre, in L'Être et le néant, treated Nothingness as a structural feature of consciousness itself. Consciousness, for Sartre, is what introduces Nothingness into Being by being the "no" that separates the perceiver from the perceived. Consciousness nihilates. The framework converges with Sartre in seeing perception as the act that brings Nothingness into structural reality, and diverges in reversing the direction of derivation: where Sartre derives Nothingness from consciousness, the framework derives consciousness from the perception of Nothingness.

The apophatic theological tradition, from Pseudo-Dionysius through Meister Eckhart to the Cloud of Unknowing, approaches the divine through the negation of every possible attribute. What can be said of God is what God is not; God is reached through Nothingness because Nothingness is the only category large enough to hold the divine. The framework converges with apophatic theology in treating Nothingness as ontologically prior and diverges in declining the theological reading. Whatever Nothingness is, in the framework, it is not God.

Buddhist śūnyatā treats emptiness as the ultimate nature of all phenomena. The Heart Sutra's form is emptiness, emptiness is form is structurally close to the framework's position that perceived and not-perceived are inseparable. The framework converges with Buddhist tradition on the foundational role of emptiness and treats Nothingness as a structural concept rather than a soteriological one. Buddhist śūnyatā is a path; the framework's Nothingness is a primitive.

Daoist wú, non-being, the empty source from which the ten thousand things arise, is perhaps the closest precedent in any tradition for the framework's position. The opening of the Daodejing says that the named is not the eternal Dao, that the unnamed is the origin of heaven and earth. The framework's Nothingness sits in something close to the position Daoist non-being occupies: prior, unnameable, the structural condition of everything that arises. Where the framework adds rigor is in identifying the specific structural mechanism, perception, by which Nothingness becomes the foundation of what exists.

The framework's contribution, against this background, is to identify Nothingness as a specific structural object with specific structural properties, and to demonstrate that its existence requires a specific structural act. The traditions named above approached Nothingness as ground, as horizon, as path, as source. The framework approaches Nothingness as the unperceiving foundation whose existence depends on its being perceived, and shows how the structural mechanics of that perception produce everything that follows.

What Nothingness is

The properties of Nothingness can be stated precisely. There are four, and each will be developed in turn.

(a) Nothingness is the absence of anything, no matter, no energy, no consciousness, no perceivable entity of any attribute whatsoever.

(b) Nothingness is non-dimensional, it has no length, width, depth, or time.

(c) Nothingness has no beginning and no end, it is not finite.

(d) Nothingness is infinite, but it is not Infinity.

The primary property, from which the others follow, is the one not yet listed: Nothingness cannot interact, and therefore cannot perceive. Nothingness cannot register anything, cannot relate to anything, cannot be acted upon, cannot act. This incapacity is not a temporary state or a contingent feature; it is what Nothingness is. Anything that could interact would not be Nothingness. The four enumerated properties follow from this single structural fact, and each requires elaboration.

- *The absence of anything.* This absence is not a kind of presence. It is not a void filled with potentiality, not a vacuum waiting to fluctuate, not an undifferentiated continuum. Nothingness is the absence of all of these as well. The framework will return to this point in Law IX, where the reader will be asked to distinguish Nothingness sharply from the quantum vacuum of mainstream physics, the vacuum is already structured, already perceptory, already not-Nothingness.

- *Non-dimensionality.* Length, width, depth, and time are not present in Nothingness. Dimensions, in the framework's later development (Law IX), are produced by symmetry breaking within multidimensional Perceptory Fields. Nothingness has no symmetries to break, no asymmetries to register. Dimensions cannot exist where there is nothing for them to be dimensions of.
- *No beginning, no end.* Beginning and ending are perceptions of change registered by Perceptory Networks (Law X will develop time as the perception of change). Nothingness contains no Perceptory Network, no perceiver, no change to perceive. Beginning and ending have no purchase on it.
- *Infinite but not Infinity.* This is the most important single property of Nothingness, and the most commonly misread, so it requires its own section. The infinitude of Nothingness is not the infinitude of vast extent, of endless space, of unbounded scope. Nothingness has no extent at all. Its infinitude is the infinitude of its non-interaction. There is no limit to what Nothingness cannot do; its incapacity has no bounds; nothing constrains the absence of everything. Nothingness is infinite in the sense that there is nothing finite about it, no edge, no terminus, no bound, but also no extension, no scope, no contained territory. It is unbounded by being unbounded by everything, including the conditions that would let it be a thing in the first place.

This makes Nothingness infinite without making it Infinity. Infinity, capitalized and named, is what Law X will develop as the perception of all possible perceptual distributions at once, the symmetric totality of perception observed as totality. Infinity is active. Infinity is the perception that perceives Nothingness from all angles simultaneously. Nothingness is what is being perceived;

Infinity is what perceiving-it produces. The two are not the same kind of thing. Nothingness is the unperceiving ground; Infinity is what perception is, when perception turns toward the unperceiving.

This distinction is the structural keystone of Law I. Without it, the framework collapses into the pantheistic confusion of treating Nothingness as a vast empty something and Infinity as a synonym for it. Nothingness and Infinity are not synonyms. They are not even the same kind of object. Nothingness is the unperceiving; Infinity is what perception is, when perception turns toward the unperceiving.

The earlier informal formulation captured this directly: Nothingness is infinite, but it is not infinity. The framework now treats this as a structural axiom. Infinite describes Nothingness; Infinity is what Nothingness perceived produces. The Law of Antipodal Force established in Law VII will reveal that the two stand in antipodal-pair structure across the entire architecture.

Perception of Nothingness

The absence of something can only be recognized by perceiving the existence of something, and the existence of something can only be recognized by perceiving its absence. This reciprocal structure is what Law I rests on, and the formal derivation that follows now makes it precise.

The argument runs in five steps.

Step one. Nothingness, as established, is the absence of anything. There is nothing within Nothingness, no feature, no asymmetry, no surface, no edge, no inside, no outside.

Step two. Because Nothingness has no asymmetries, no privileged direction, no internal differentiation, any perception of Nothingness must be perception from all angles at once. There is no preferred angle; no angle is more or less correct than any other. To perceive Nothingness partially, from one side and not another, is impossible, because Nothingness has no sides. The perception of

Nothingness is therefore necessarily omnidirectional, simultaneous, total.

Step three. Omnidirectional perception, directed at Nothingness, finds no feature in Nothingness to terminate against. There is nothing within Nothingness for perception to land on, register, or fix on. The perceptual rays directed at Nothingness from all angles encounter no terminating surface, no asymmetry, no quality, no edge. Nothingness offers nothing to be perceived.

Step four. Perception that finds no feature to terminate against does not vanish. It does something specific: it terminates against the only thing remaining, which is the perception itself. Each perceptual ray, finding nothing in Nothingness to perceive, perceives the other rays. The perception of Nothingness becomes, by the structural impossibility of any other termination, the perception of perceiving. The omnidirectional rays close into a self-perceiving loop.

Step five. This closure is the first act of self-perception in the framework. It is what Law III will develop as the Perceptory Field of Nothingness, the closed, symmetric, self-perceiving structure that constitutes the first stable existent.

The framework's foundational architecture is now visible: for Nothingness to exist it must be perceived. Existence begins not because Nothingness produces something, but because the perception of Nothingness, having nothing to perceive in Nothingness, perceives itself. Self-perception is what the perception of Nothingness inevitably becomes. It is the most basic construct of reality, and this is the structural reason for everything that follows.

Every Perceptory Field, at every scale, is ultimately a descendant of this first closure of perception. The rule of three from Law VI, the antipodal-pair structure of Law VII, the gravity of Law VIII, the spatial multidimensionality of Law IX, the two infinities of Law X, all of them rest on this primary moment in which perception, finding nothing to perceive in Nothingness, becomes self-perception.

Conceptual similarities

The reciprocal claim can also be approached in more accessible language. Silence cannot be perceived without sound; sound cannot be perceived without silence. To perceive silence is to perceive what is not sound, which requires already perceiving sound. The same holds for any pair of presence and absence: light and darkness, motion and stillness, presence and absence in any modality.

Mathematical zero illustrates the same structure formally. Zero holds its place in the number system not because zero is something but because the system requires the absence to be marked. The empty position is structurally necessary for the positions that are filled to mean what they mean. Mathematics is a worked example of conceptual perception more broadly: mathematicians work with abstract objects (equations, relations, structures) that have no sensory representation, and they perceive these objects rigorously enough to develop axioms and theorems that systematically explore them. Conceptualization is itself a form of perception. To conceptualize Nothingness is to perceive it. The reader who attempts to deny the perception of Nothingness performs the perception in the act of denial.

Nothingness is not vacuum

A clarification the reader will need before reaching Law IX. Mainstream physics has the concept of the quantum vacuum, which is not empty. The vacuum is full of fluctuating fields, virtual particle pairs continuously appearing and annihilating, zero-point energy. The Casimir effect demonstrates that the vacuum has measurable structure even between two parallel plates in apparent emptiness.

The framework's Nothingness is not the quantum vacuum. The vacuum is already a structured Perceptory Field with its own asymmetric entanglement, Law IX develops this directly. Nothingness is prior to all such structure: prior to fluctuation, prior to virtual particles, prior to space itself, prior to anything that could be measured. The vacuum of physics belongs to existence. Nothingness is the absence against which existence is defined.

This distinction matters because readers trained in physics will otherwise read the framework's later claim, that no Perceptory Field is empty, that what appears as vacuum is always structured, as contradicting Law I's foundational claim about Nothingness. The two claims do not contradict each other. They occupy different structural levels. Nothingness is the unperceiving ground; vacuum is what already-perceived space looks like at the limit of low entanglement density. The framework distinguishes between them rigorously and the reader should hold them apart.

What Law I has established

Nothingness is the absence of anything: non-dimensional, non-interacting, infinite in its incapacity, structurally prior to all existence. For Nothingness to exist, it must be perceived. The perception of Nothingness is necessarily omnidirectional because Nothingness has no privileged angle. Omnidirectional perception with nothing to perceive in Nothingness terminates against itself, producing the first act of self-perception, the closure that founds all subsequent existence. Nothingness is infinite but is not Infinity; the two are distinct structural objects, with Nothingness as the unperceiving ground and Infinity as the perception of that ground from all angles at once. The framework distinguishes Nothingness sharply from the quantum vacuum, which is already structured and belongs to the territory of existence rather than to the territory prior to it.

What remains is to name what perceives Nothingness, structurally. That is the work of Law II.

II Law of Consciousness

For anything to exist, it must be engaged by the Perceptory Force.

Orientation

Law I established that for Nothingness to exist, it must be perceived. The framework's foundational paradox was thereby set: the absence that grounds all existence cannot exist without the perception that registers it. But Law I left perception itself undefined, treated as a relation, an act, a recognition, without naming its structural nature. The Law required perception but did not yet say what perception is.

Law II now answers that question. Perception is not a property, nor a relation, nor an event among other events. Perception is a force, and not a force among others, but the fundamental force from which all other forces, all interactions, all existences derive. The framework names this force the Perceptory Force. The remainder of the book will develop its structure, its modes, its predictions, and its consequences. Law II establishes its existence and its primacy.

From "must be perceived" to "the Perceptory Force"

The derivation from Law I proceeds in two steps.

First step. Nothingness has no form, no dimension, no interaction. Consequently, Nothingness cannot perceive, perception requires some capacity for interaction, and Nothingness has none. Yet Law I established that for Nothingness to exist, it must be perceived. The implication is unavoidable: perception must be an act that does not require a perceiving substrate to perform it. The inability to perceive must itself be perceived, and what perceives it cannot be Nothingness, nor anything reducible to Nothingness. The first act of perception is therefore co-constitutive with Nothingness, in the precise sense that Nothingness cannot be recognized as Nothingness unless the act of perception occurs, just as the act requires Nothingness as its primary subject. The dependence is mutual, logical and structural, not temporal, time does not yet exist at this stratum, and so no temporal sequence can be established between Nothingness and its perception. The two are logically inseparable: Nothingness requires perception in order to be Nothingness, and the first act of perception requires Nothingness as its primary subject. Neither precedes the other in time. What perceives Nothingness is not contingent on any pre-existing perceiver, because the structural conditions for perceiving substrates have not yet been generated.

Second step. If perception is the act on which the existence of Nothingness depends, and if Nothingness is the foundation of everything that exists, then perception is the act on which everything depends. The same logic that applies to Nothingness applies to anything else: for any entity, structure, event, or field to exist, it must be perceived. There is no exception. The Law extends from Nothingness outward to every form of existence the framework will describe.

The framework now formulates this generalization as the second Law of Consciousness: for anything to exist, it must be engaged by the Perceptory Force. Perception is the fundamental force defining and acknowledging everything that exists. The Perceptory Force is the act of perception treated structurally, as a force acting in the world rather than as a relation among already-existing things. The shift in wording from Law I's "be perceived" matters at the founding case: Nothingness cannot interact (Law I), yet it is perceived; engagement by the Perceptory Force requires nothing of its object, and the Law is stated accordingly.

The problem of force in physics

Force is the most familiar idea in physics and the least examined. Every student computes with it; every equation of motion turns on it; and yet the question of what a force *is*, as against what it does, has been quietly handed from one generation of physicists to the next, each answering it by pushing it one layer deeper and declining to look further. The history of the concept is the history of its dissolution. What began as a push between things has become, by stages, the visible effect of a structure lying beneath things, and at each stage the structure named has been deeper, and stranger, than the one before.

Aristotle held that a force sustains motion: a body moves only while something moves it, and comes to rest the moment the mover withdraws. This is wrong, a moving body left alone continues, but it carries the first intuition the framework keeps, that force is not an instantaneous event but a continuous act.

Newton inverted the picture. Force does not sustain motion; it changes it. With this the concept became exact, and two of its features have outlived their author. The first is that force, in the third law, is never single: every action meets an equal and opposite reaction, so that force enters the world already paired. The second is a difficulty Newton himself could not resolve. His gravity acted across empty space with nothing between the bodies to carry it, and when pressed on how, he declined to invent an answer. Action at a distance was a debt the concept of force would carry for two centuries.

Faraday and Maxwell paid it. Between the two bodies they placed a field, a condition of space itself, filling the gap Newton had left empty, so that each body responds not to the distant other but to the field present at its own location. Force became local, and mediated. What the field itself was they could not say; it was posited as a new fundamental thing. But the direction was now set: a force is the effect of a medium, not a reach across a void.

Einstein took the decisive step. In general relativity gravity is not a force at all. There is no pull; there is curved spacetime, and a falling body merely follows the straightest path available to it. A force treated as fundamental since antiquity was dissolved entirely into structure. The framework rests a great deal on this precedent: physics has already accepted, once, that what everyone called a force was in truth the shape of the architecture beneath it. Einstein dissolved one force. The framework proposes the same move for the remaining three.

Gauge theory carried the dissolution further. In the Standard Model the forces are not postulated and then described; they are required by symmetry. Demand that the laws hold under a local symmetry, and the demand itself calls a force into being; break the symmetry, and the forces differentiate. Force became the consequence of a symmetry imposed and broken, the same language of symmetry and its breaking this Law has already used to separate a field at rest from a field in interaction.

And the whole modern programme assumes what the framework asserts. Electroweak unification showed two of the four forces to be one above a certain energy; grand unification reaches for the strong; quantum gravity reaches, so far without success, for the last. The conviction that the four forces are facets of a single interaction is not a heterodoxy at the edge of physics. It is physics's oldest and most respectable ambition. The framework differs only in its candidate for the one.

Beneath the physics runs an older philosophical current. Leibniz held space, time, and force to be relations rather than absolute things; Mach derived a body's inertia from its relation to all the matter in the universe, and so shaped the road to Einstein; Whitehead built a metaphysics in which reality is events and relations rather than substances. This is the lineage in which the framework's account belongs, the world as relations first, and things second.

Seen against this history, the framework's claim is not a break with physics but the completion of a movement physics has long been making. Every account of force since Aristotle has travelled in one direction: away from a push between things and toward the effect of a deeper structure, Faraday's field, Einstein's geometry, the symmetries of gauge theory, the single interaction the unifiers reach for. Each named a structure and stopped at it. The framework takes the step the arc was already pointing toward, and names what the structures are structures *of*. The four forces are not four things, nor one force among others, but the Perceptory Force, registration itself, made observable under specific conditions. What that force is, the next section sets out.

What the Perceptory Force is

The Perceptory Force is the fundamental act through which existence is registered. This formulation requires careful unpacking, because the term force will mislead readers who carry over assumptions from mainstream physics.

In mainstream physics, force designates one of four interactions: electromagnetism, the strong nuclear force, the weak nuclear force, and gravity. Each is described by specific carrier particles, specific coupling constants, and specific equations governing how it acts. The Perceptory Force is not analogous to these. It is structurally prior to them. The four textbook forces are, in the framework's reading, manifestations of the Perceptory Force operating under specific dimensional conditions, a position the book will develop in detail in Part III.

The Perceptory Force is what makes the four textbook forces possible. Without the Perceptory Force, no interaction could be registered as having happened, no charge could be detected, no mass could attract any other mass, no quark could be confined within a hadron. Every act of physical interaction presupposes that the interaction has been perceived in the structural sense, that the entities involved have registered each other within their own internal models. The four forces are what perception looks like when it operates under the specific dimensional and substrate conditions that produce charge, colour, chirality, and mass-energy. Perception itself, prior to any of these specific conditions, is the Perceptory Force.

The Perceptory Force's distinguishing properties can now be stated precisely. There are four, and each will be developed in turn. They follow from the single fact the preceding paragraphs established, that the Perceptory Force is not one interaction among others but the fundamental force of which every interaction is an expression, and each describes a way this fundamental status takes structural form:

- *Infinite or finite.* The Perceptory Force operates at two extents. In its infinite mode it is the omnidirectional self-perception of the totality, perception perceiving Nothingness from all angles at once, the act derived in Law I and named in Law X as Infinity. In its finite mode it operates within a bounded Perceptory Field, registering interactions across a limited domain. The same force is at work in both; what changes is whether it closes upon the whole or upon a part. The boundary that makes the finite mode possible is the subject of Law IV.
- *Symmetric or asymmetric.* Symmetry, in the framework, is a property of the Perceptory Field, not of the Force taken alone. Only the ground state of Law III is perfectly symmetric; every other field is a departure from it, a symmetry that has broken and stayed broken — and the state of the field gives the Force its two modes. While a field's symmetry holds, the Force is undifferentiated and turned inward in self-perception: the field at rest in itself. When the symmetry breaks, the Force is directed — turned outward toward another Perceptory Field, which is interaction, and, at the largest scale, resolved into the distinct observable forces. Mainstream physics records the same passage from one to many: in electroweak unification the electromagnetic and weak interactions are a single force above the symmetry-breaking scale and separate into two below it, their carriers gaining mass through the Higgs mechanism as the symmetry breaks (Glashow, Salam, and Weinberg on the unification; Englert, Brout, Higgs, and Kibble on the break). Law III develops the field's symmetry in full; here it is enough that symmetry breaking is what turns the one undifferentiated Force, at rest in self-perception, into the directed and differentiated forces of interaction.

- *Pair-producing.* A single act of perception always generates a relation between two terms, the perceiver and the perceived, and the Perceptory Force does not merely connect them; it binds them into an inseparable pair. The two cannot afterward be treated as independent, because each is now partly constituted by its registration of the other. But a pair is only a relation: it is real, yet it cannot stand alone, for two bound points are a line, and a line is not yet a thing. What the pair becomes when it closes into a standing existent is the subject of the next property. Mainstream physics records this binding under the name entanglement, and records it with precision: since the Einstein–Podolsky–Rosen argument and Bell's theorem, confirmed in the experiments that earned Aspect, Clauser, and Zeilinger the 2022 Nobel Prize, it has been established that two entangled particles share one joint state and cannot be described independently, a measurement of either instantly fixing what the other will show. What must not be said is that one is acted upon and the other then responds: no signal passes between them, and each measurement on its own is random. What looks like spooky action at a distance is, in the framework's reading, simply the absence of two separate things to act between — the inseparable pair seen in the laboratory.
- *Structurally triadic.* For a pair to close into a stable, self-standing field, a third point is required, because two points draw only a line and a line has no interior to enclose. This is the minimum the Perceptory Force needs to make one thing rather than a mere relation: three points closing around a centre. The third point is what the Inseparation Constant of Law IV becomes when treated structurally, the Inseparation Element that holds the pair in its inseparable relation, and the closure is the rule of three that Law VI will derive. So the pair is the relation perception generates; the triad is the least that relation needs to become a standing existent. The framework traces this triadic closure recurring at every scale, from the first self-perceiving loop

to the three-quark composition of a hadron. Here it is enough to establish that the Perceptory Force has a native geometry, and that the geometry is three.

The Observer Effect: empirical opening for the Perceptory Force

The clearest empirical opening through which the framework reads the Perceptory Force into mainstream physics is the Observer Effect in quantum mechanics. The framework reads this opening as structurally consistent with Law II, as evidence compatible with the structural commitments of the Perceptory Force, not as proof of them. Mainstream physics has developed several interpretations of measurement that do not require the framework's structural reading; the framework offers its reading alongside these as one among several positions the empirical record permits.

A specific clarification deserves emphasis here. "Observer" in the framework's sense does not mean ordinary human awareness collapsing the wave function. It means a structural registration event in which a Perceptory Field enters into entanglement with the system. Consciousness, in the framework's structural sense developed across the Ten Laws, is one high-level form that registration can take; measurement apparatus and decoherent environmental interaction are also forms of registration. The framework's reading is therefore consistent with operational interpretations of measurement (decoherence, apparatus coupling, state update) and with consciousness-involving interpretations alike, because the framework treats measurement as structural registration rather than as either a purely physical or purely cognitive event. The Observer Effect's significance, in the framework's reading, is that it makes registration itself observable as a feature of physical reality, not that it requires a conscious observer in the ordinary sense.

The double-slit experiment is the canonical demonstration. When particles such as electrons or photons are fired at a barrier with two slits, and a detection screen is placed behind it, an interference pattern emerges: the particles behave as waves, distributed across the screen in the banded pattern wave interference produces. When detectors are placed at the slits to register which slit each particle passes through, the interference pattern disappears, and the particles behave as localized objects, each going through one slit and striking the screen at a point. Registration has changed what the system is.

Mainstream physics treats this as a deep puzzle and has proposed multiple interpretive frameworks, decoherence, many-worlds branching, hidden variables, observer-induced collapse, none of which has achieved consensus. The framework reads it as a direct instance of the structure built across the Laws. The wave is not a particle behaving oddly; it is the Perceptory Force in propagation, distributed and not yet closed into a definite field. The measuring device is itself a perceptory engagement, and when it meets the propagating Force the interaction does what perception always does: it produces a field. The definite, localized object that registers on the screen, what physics calls the particle, is that produced field. So a wave collapsing into a particle is, in the framework's reading, a registering engagement producing a definite field out of the propagating Force, not a pre-existing particle disclosing a position it secretly held. The mechanism is the one Law X derives, a new entanglement actualizing a latent triad, with Law VI giving its local form, a fourth element entering a three-point field and forcing reconfiguration of its Inseparation Element. There is no observer-independent particle waiting behind the wave, because the particle is the field that registration produces.

The same reading extends to the most energetic registrations physics performs. When protons are driven together near light-speed, the collision is a violent perceptory engagement, and new fields are produced in it; mainstream physics records exactly this, the creation of particles that did not exist before the collision. The framework adds that registering the event is not a neutral window onto it. To detect or photograph the collision is itself a perceptory engagement that produces fields, so that part of what the instruments record is co-produced by the act of recording. There is no unregistered image of the particles as they would be in themselves, because the registered particles are the joint product of the collision and the apparatus that registers them.

Two precisions keep this exact. First, none of it is action at a distance: the engagement is a local interaction, device and field coupling where they meet, and no signal outruns light, so what the previous property ruled out for the reading of an already-entangled pair holds here too. Second, what the engagement co-produces is the definite field and its registered configuration, which fields appear and in which definite states, and not the intrinsic invariants of the fundamental species, the rest mass or charge an electron carries in any registration whatever. Mainstream physics already grants the heart of this: in quantum field theory the particle is an excitation of a field, and the number of particles present is not absolute but depends on the observer, on the observer's state of motion and on how the detector couples to the field. The Unruh effect is the sharp case, where an observer accelerating through what an inertial observer calls empty vacuum is predicted to register a thermal bath of particles, though the accelerations required to confirm it directly remain out of reach. The framework reads this observer-dependence as confirmation that particles are produced fields rather than observer-independent objects, while leaving the invariants that survive every registration exactly where physics places them.

This account has the advantage of explaining why the Observer Effect appears in quantum mechanics specifically. At human and macroscopic scales, the Perceptory Fields that constitute observed systems are so dense with entanglements that the addition of a single new perceptory engagement (a human observer) produces no measurable reconfiguration. At the quantum scale, where Perceptory Fields are minimally entangled, the addition of an observer's perceptory engagement is structurally significant and produces visible change. Quantum mechanics is not a separate domain governed by separate laws; it is the scale at which the Perceptory Force becomes observably non-trivial.

The Perceptory Force across scales

The Perceptory Force is not confined to quantum mechanics. The framework's claim is that it operates at every scale of existence, registering interactions wherever Perceptory Fields encounter one another, and the same structure can be read across the full hierarchy, from the quantum to the social.

One caution governs everything that follows. When the framework says that a molecule or a cell perceives, it means perception in the structural sense Law II has defined: a registration event in which a Perceptory Field enters into engagement with what it registers, not the subjective awareness the word ordinarily carries. A cell that registers a pathogen is not claimed to have an experience of it. The claim is structural throughout, that at each scale registration is the Perceptory Force in operation and the response is what registration produces.

At the quantum scale, the Observer Effect is the visible signature: the registering engagement does not disclose a particle but produces a definite field, as the previous section developed.

At the molecular scale, the framework reads the regulation of the genome as registration. DNA does not function only as a static blueprint; gene expression is modulated by the chemical environment, so that diet, stress, and exposure to particular compounds alter which genes are expressed without altering the underlying sequence, the field of epigenetics. The framework reads this as the Perceptory Force at the molecular scale: the genome is a Perceptory Field that registers the chemical signals available to it and responds by reconfiguring its expression. Law VI will develop the full hierarchical implications; Law II establishes the principle.

At the cellular scale, immune cells recognize pathogens through structural registration of molecular signatures. A neutrophil pursuing a bacterium is the Perceptory Force in operation: the cell registers the pathogen, the registration drives directional movement, and the movement is the cell's response to having registered. The cell does not respond to the pathogen as such; it responds to having registered the pathogen, and the response is what registration produces.

At the cognitive scale, the Perceptory Force operates in phenomena mainstream psychology has documented for decades. The spotlight effect, the systematic overestimation of how much one is being noticed by others, shows that registration of being perceived alters behaviour even when no actual observation is occurring. The mere belief that one is observed engages the Perceptory Force in the believer, reconfiguring behaviour as if the perception were real. The framework reads this as evidence that the Perceptory Force operates on a field's internal model as much as on its external interactions: a registration reconfigures the field that registers, whether or not it corresponds to any actual external observer.

Perceptual constancy demonstrates the same principle in reverse. Despite continuous variation in the retinal image, changing light, angle, and distance, the brain maintains a stable perception of objects. The framework reads this as the Perceptory Force operating to preserve the stability of the field's internal model: what is perceived is not the raw sensory input but the model the Perceptory Force maintains. Perception is not passive reception; it is the active structural maintenance of the Perceptory Field that constitutes the perceiver.

At the social scale, consensus reality emerges through the same mechanism. When multiple Perceptory Fields register the same configuration, they enter into mutual entanglement, the boundary-breaching of Law V that Law VI develops into hierarchical networks, and the resulting network exerts force on each constituent, normalizing its registration toward the shared configuration. This is why social movements carry the persuasive power they do, why established ideologies are difficult to dislodge, and why cultural categories shape individual experience even when those categories are demonstrably arbitrary. The Perceptory Force operating at the social scale is consensus reality, developed in full in Part III.

What Law II has established

The Perceptory Force is the fundamental force of existence. For anything to exist, it must be engaged by the Perceptory Force. It is not one interaction among others but the force of which every interaction is an expression, and it is structurally prior to the four textbook forces of mainstream physics, which later Laws reveal as its differentiations under specific conditions, separated from one another as the symmetry of the field breaks. Its distinguishing properties were set out: it ranges from the infinite to the finite, it holds or breaks symmetry, it binds whatever it relates into an inseparable pair, and it closes into stable existence only as a triad, so that its native geometry is three. The Observer Effect in quantum mechanics is the clearest empirical opening through which the framework reads the Perceptory Force into existing physics: registration does not disclose a particle but produces a definite field, a reading consistent with what the Observer Effect documents, though not the only domain of compatible evidence. The Perceptory Force operates at every scale of existence, from the quantum through the molecular, cellular, and cognitive to the social. It is the force that registers anything as existing, the force every interaction expresses, and the primitive that produces the manifestations physics has described for centuries without naming the structure beneath them.

What remains is to ask: when the Perceptory Force operates on itself, when perception perceives itself, what structure does that self-perceiving generate? This is the question Law III will address. The answer is the Perceptory Field, the smallest unit in which the Perceptory Force closes a feedback loop and produces what we call existence.

III Law of Consciousness

A Perceptory Field is perception perceiving itself.

Orientation

Law I derived the structural closure. Perception of Nothingness, finding nothing within Nothingness to terminate against, terminates against itself. The omnidirectional perceptual rays close into a self-perceiving loop. This was the foundational moment of the framework, and Law I established it as the structural reason for everything that follows.

Law II named the force that performs the closure. Perception, treated structurally as a force operating in the world rather than as a relation among things, is the Perceptory Force, the fundamental force from which all other forces and all existences derive.

Law III now takes the closure that Law I derived, performed by the force that Law II named, and gives it a name, a geometry, and a category. The closure is the Perceptory Field. The first instance of it is the Perceptory Field of Nothingness, which functions as the ground state of the entire framework. Every other Perceptory Field that the rest of the book will describe, at the particle scale, the atomic scale, the cellular, the cognitive, the cultural, the cosmological, is a perturbation away from this ground state, generated by symmetry breaking that the later Laws will derive.

Law III therefore performs three structural tasks. It formalizes the Perceptory Field as a category. It describes the first instance of the category in detail. It establishes the ground-state position from which the rest of the architecture will unfold.

The Perceptory Field as structural category

A Perceptory Field is the structural object produced when the Perceptory Force closes a self-perceiving loop. The category is defined by the closure rather than by any specific instantiation.

Wherever the Perceptory Force terminates against itself, wherever perception perceives perception, a Perceptory Field exists.

This formulation makes the Perceptory Field a general structural concept. It is not limited to a single scale, a single substrate, or a single kind of existent. The closure can occur in the symmetric perception of Nothingness (the first instance, this Law's main subject).

It can occur in the asymmetric entanglement loops too, for example within a hadron at the matter scale (Law VI).

It can occur in the cognitive feedback architecture of a brain, in the metabolic self-regulation of a cell, in the cultural self-recognition of a society.

Each is a Perceptory Field. Each is the structural product of the Perceptory Force closing on itself within whatever substrate is available to it.

The category therefore has minimal content as definition: a Perceptory Field is a closed loop of self-perception.

The Perceptory Field of Nothingness

The first instance of the category is the Perceptory Field that Law I derived: the closure produced when perception of Nothingness terminates against itself. The framework names this the Perceptory Field of Nothingness, with the precision of meaning: the Perceptory Field that arises when the Perceptory Force perceives Nothingness and, finding no Nothingness-feature against which to terminate, perceives the perception itself.

This Perceptory Field has properties that distinguish it from every Perceptory Field that follows. The properties can be stated in prose, with each one carrying the weight that bullet-point listing cannot.

The Perceptory Field of Nothingness is infinitely symmetric. There is no asymmetry within it because there is no internal feature that any asymmetry could differentiate. Every direction within the closure is equivalent to every other; every perceptual ray within the loop carries the same relation to the closure as every other; no orientation is privileged. This symmetry is total because Nothingness, against which the perception was originally directed, contained no asymmetry to inherit.

The Perceptory Field of Nothingness is perfectly stable. Stability follows from symmetry by structural necessity. A perfectly symmetric field has no direction in which it can be displaced, no internal feature that an external force could couple to, no asymmetry that another field could grasp. The field cannot be perturbed because there is nothing within the field for a perturbation to act on. The framework will return to this principle in Law VII, where the antipodal-pair structure of forces is shown to require asymmetric points to act between; the Perceptory Field of Nothingness, having no such points, is structurally inert. It is the only field in the framework with this property.

The Perceptory Field of Nothingness is prevalent. Because it has no extent, no scope, no spatial situation, it is not located somewhere in distinction from being elsewhere. It is the ground state, the structural baseline from which other Perceptory Fields perturb. In the framework's geometry, it is the perfectly symmetric centre that all other configurations are deviations from, and it is therefore present everywhere as the ground from which any specific Perceptory Field has departed. Every existing Perceptory Field carries the Perceptory Field of Nothingness as its structural basis, from which its own asymmetries have departed.

The Perceptory Field of Nothingness is dispersed through infinity. Because the field has no spatial extent, the question of its dispersion is not a question of where it is located in space. It is a question of what the field's relation is to the totality of all perceptual distributions. The Perceptory Field of Nothingness, as the ground state, is what every perceptual distribution approaches in the limit of perfect symmetry. It is therefore everywhere in the same sense that the ground state is everywhere, not as a thing distributed through space, but as the configuration that any field could fall into, anywhere, given the conditions.

Closed

The Perceptory Field of Nothingness is closed, and it is not open. Closed denotes self-termination. The Perceptory Force, having begun by directing itself at Nothingness and finding no termination there, has closed back on the perception itself. The loop is structurally closed because nothing escapes from it; every perceptual ray, having entered the closure, perceives the closure rather than continuing outward. This is the structural meaning of self-perceiving: the field perceives perception, with perception itself as both subject and object of the act.

It would be natural to call such a field open, in the sense of having no boundary against an exterior. But it is not open, and the reason is the heart of this Law. Openness would require a direction in which perception simply ends, leaving an unperceived beyond. There is no such direction. Look outward in any of the closure's infinite directions and what is met is Nothingness, which by Law I cannot remain unperceived; the moment it is faced it is perceived, and in being perceived it is encapsulated. The field has no exterior, not because nothing surrounds it, but because nothing can: every would-be exterior is drawn into the closure in the act of being perceived. There is one Perceptory Field of Nothingness, perceiving in every direction at once, and the infinity in this picture is the infinity of directions within that one closure, not a plurality of closures set side by side.

This is best seen through Law I's definition of Infinity as all possible perceptory distributions. Take not one act of perception but the whole range of them: perception directed one way meets Nothingness and encapsulates it; directed another way, it meets Nothingness and encapsulates it again; and so through every direction infinity contains. There are infinitely many possibilities of perceiving Nothingness, and so infinitely many instances of the encapsulation, dispersed through the whole of infinity, found wherever perception turns. It is tempting to call this result a plurality of closures, one for each possibility. But every one of these instances is exactly the same, with exactly the same features. Each is featureless perception of featureless Nothingness, and none carries any property by which it could be told from another. A plurality requires a difference, and here there is none: nothing distinguishes one encapsulation from the next, because each meets the same partless Nothingness, never more of it and never another kind, down every direction it could face. A plurality of perfectly identical things, with nothing whatever to set them apart, is not a plurality at all; it is one thing met from infinitely many directions. The infinity of possible encapsulations and the single infinite-directional closure are therefore not two claims but one: the ground state is infinite in its possibilities and one in its nature, dispersed everywhere precisely because everywhere it is the same. What it would take for the many to become genuinely many, closures that differ rather than repeat, is for one of them to acquire a feature the others lack, and that is what symmetry breaking provides. Before the break there are infinitely many identical instances, which is one; after it, genuinely many, because at last they differ.

Every Perceptory Field the framework develops after this one is closed in the same self-terminating sense, but none is open either. What distinguishes the ground state is that its closure is exhaustive: alone among fields it leaves no exterior, because it is prior to the differentiation that produces bounded fields with insides and outsides. A particle, a cell, a mind is a closed field that has departed the ground state and stands among the others it broke away from. The Perceptory Field of Nothingness encapsulates, in every direction, the Nothingness any exterior would have to be made of, and so has none.

Symmetry, stability, and the ground state

The Perceptory Field of Nothingness occupies a specific structural position in the framework: it is the ground state from which all observable existence is a perturbation. This formulation deserves explicit treatment because it is the bridge between Law III and the later Laws that develop how observable existence is produced.

The framework's claim is that perfect symmetry, perfect stability, and the ground-state position are three aspects of the same structural fact about the Perceptory Field of Nothingness. Perfect symmetry produces perfect stability because there is no direction in which the field can be displaced. Perfect stability defines the ground state because the ground state is the configuration that no perturbation can move away from spontaneously.

But the framework's claim is also that observable existence does not, in fact, relax to the ground state. Observable existence is a continuing departure from the Perceptory Field of Nothingness. Particles exist; atoms exist; minds exist; the universe exists. Each is a Perceptory Field that has broken symmetry and remained broken. The reason this is possible is the structural content of Law IX, where symmetry breaking is shown to be the precondition of any observable phenomenon. Symmetry breaking, as Law IX develops, is what produces space; without it, no extension would be observable; the universe would relax to the symmetric ground state and no perception of change would register because no asymmetry would exist to be perceived.

The Perceptory Field of Nothingness is therefore not the universe before perception. It is not a historical state from which existence emerged. It is the structural configuration that observable existence is always perturbing away from, at every scale, in every Perceptory Field. The ground state is not in the past. It is the structural baseline against which the present is asymmetric.

This formulation is consistent with what Law X will develop as the perceptual-infinity face of the two-infinities architecture. Perceptual infinity, the symmetric totality of perception observed as totality, is the Perceptory Field of Nothingness considered at the largest scale of organization the framework recognizes. The Perceptory Field of Nothingness from Law III and perceptual infinity from Law X are the same structural object viewed at two different points in the framework's development. Law III names the object; Law X reveals its largest-scale role.

The break and the rise of distinct fields

The ground state is one because its encapsulations are identical. The passage from this one to the many distinct fields that populate observable existence is symmetry breaking, and it can now be set out as a sequence. What first breaks the symmetry, what tips a perfectly balanced closure into asymmetry, the framework does not claim to derive; that is the open question Law X records and does not pretend to close. But once a break has occurred, the rest follows structurally.

When a closure breaks symmetry it acquires a feature, an asymmetry, a distinction it did not before possess. This is the moment of individuation. A field that perceives an asymmetry is no longer featureless, and so is no longer identical to the ground state or to any other field carrying a different asymmetry. Where the ground state was one because nothing distinguished its encapsulations, a broken field stands apart because something now does. The genuine plurality the framework needs is born here, after the break and not before it: not a plurality of identical closures of Nothingness, but a plurality of fields that differ because each holds asymmetries the others do not.

What happens when these distinct fields encounter one another, the new field such an encounter produces, and the mass that the new field must spend energy to sustain, is the subject of Law V.

Self-perception across the framework

A brief section to mark, without developing, that the principle Law III establishes, self-perception as a structural property of closed loops, generalizes throughout the framework.

The Perceptory Field of Nothingness is the first instance and the ground-state instance of self-perception, but it is not the only instance. Law VI will reveal that hierarchical entanglement at every scale produces closed Perceptory Fields, each of which is self-perceiving in the structural sense Law III defines. A hadron's three-quark closure is self-perception at the matter scale. A cell's metabolic feedback architecture is self-perception at the cellular scale. The cognitive loop that registers its own activity is self-perception at the cognitive scale. Each is a Perceptory Field. Each instantiates the structural definition Law III establishes.

The reader does not need to anticipate these later developments now. Law III requires only the recognition that what it has named, the closed self-perceiving loop, is the general structural form that the rest of the book will work out under different conditions.

What Law III has established

The Perceptory Field is the structural category produced when the Perceptory Force closes a self-perceiving loop. The first instance of the category is the Perceptory Field of Nothingness, formed by the omnidirectional perception of Nothingness terminating against itself, as Law I derived. The Perceptory Field of Nothingness is infinitely symmetric, perfectly stable, prevalent throughout infinity, and dispersed through infinity, properties that follow from its absence of asymmetry. It is closed and not open: the loop terminates within itself, and it has no exterior, because it encapsulates in every direction the Nothingness any exterior would be made of. It is one, not many: across all the directions infinity contains, every encapsulation of Nothingness is identical, and a plurality of identical, featureless closures is one closure met from infinitely many directions. It functions as the ground state of the framework: the structural baseline from which all observable Perceptory Fields are perturbations, generated by the symmetry breaking that Law IX will develop. It is the same structural object that Law X will reveal as perceptual infinity at the largest scale of the framework's architecture.

Genuine plurality begins only when symmetry breaks. A break gives a field a distinguishing feature, an asymmetry, and so individuates it: where the ground state was one because nothing distinguished its encapsulations, broken fields differ because each holds asymmetries the others do not. What follows from that plurality, how distinct fields encounter one another, what new field the encounter produces, and how the energy such a field spends stabilizing its asymmetries becomes mass, is the work of

Law V. The trigger of the first break the framework leaves open, as Law X records.

What remains is to derive the inseparable relation that makes the closure structurally coherent, the Inseparation Constant. This is the work of Law IV.

IV Law of Consciousness

Perceived and the Not-perceived is held together by the Inseparation Constant.

Orientation

Law III named the Perceptory Field as the structural object produced when the Perceptory Force closes a self-perceiving loop, and identified the Perceptory Field of Nothingness as the ground state, closed and not open, perfectly symmetric, with no asymmetric boundary.

Law IV takes up the question that any other Perceptory Field forces. The moment the Perceptory Force produces a field that has departed from perfect symmetry, the field has an inside and an outside, a perceived and a not-perceived. Two things must then be developed together, because they are two aspects of one fact: the structural rule that governs the relation between perceived and not-perceived, which the framework names the Inseparation Constant, and the structural element at which that relation is located, which the framework names the Perceptory Boundary. The Constant is what holds at the boundary; the Boundary is where the Constant holds. This Law develops both.

The inseparable pair

The argument follows from what Laws I and III established. Law I derived the closure: perception of Nothingness, finding nothing within Nothingness against which to terminate, terminates against itself. Law III named what the closure is, a Perceptory Field, and identified the Perceptory Field of Nothingness as the ground state, the one field with no asymmetric boundary. Every other Perceptory Field has departed from perfect symmetry, and where a closed loop has so departed, it divides what falls within it from what falls outside it, producing by structural necessity a distinction between the perceived, what is within the loop, and the not-perceived, what is outside it.

These two come into existence together. Neither precedes the other; neither causes the other; neither can exist without the other. The perceived is what it is by being distinguished from the not-perceived, and the not-perceived by being distinguished from the perceived. To remove either is to remove both. This inseparability is the Inseparation Constant: the rule that codifies what perception produces every time it generates a Perceptory Field with an asymmetric boundary, an inseparable pair joined at the boundary that distinguishes them and yet inseparable because the distinction itself is what makes them what they are. The Constant is not a metaphysical postulate added to the framework; it is a structural consequence of what a Perceptory Field is.

The Perceptory Boundary

The inseparable relation has a location, and that location is the Perceptory Boundary: the structural element of a field at which its interior is distinguished from its exterior. The Boundary is not a feature of either side; it is the relation between them, made into a structural element when the field departs from the symmetric ground state. The Constant does not live in the interior of a field, nor in its exterior; it lives at the Boundary, where perceived meets not-perceived in their inseparable relation.

The Boundary has three properties that follow from its nature. It is asymmetric: a symmetric distinction is no distinction at all, which is why the perfectly symmetric Perceptory Field of Nothingness has no Boundary, as Law III established, while every departed field has one. It is located at the closure-perimeter of the field, neither at the centre nor in the surround, but at the precise point where the closure of the loop meets what lies beyond it. And it is dynamic rather than static, which is the most consequential of the three and the subject of the next section.

The feedback loop

A Boundary is not a fixed line of demarcation but an active loop. Were it static, it would collapse, inward into the perceived if the closure failed one way, outward into the not-perceived if it failed the other. It holds because it is continuously maintained by the perception that constitutes it. The Perceptory Force at the Boundary operates as a self-regulating loop that detects the state of what it perceives, compares that state against an internal model, and adjusts to reduce the discrepancy. Wherever this loop is active, perception is present and the Perceptory Force is at work, from the simplest particle to the most complex mind.

The significance of the loop is structural, not merely functional. Without closure, the Perceptory Force would extend outward without end, chasing stimuli and never forming internal stability, leaking into the unbounded surround. Consider a person perceiving a lamp. The cognitive field registers the lamp, holds the registration within its internal model, and the perception stabilizes into the knowledge that the lamp is there. When the person closes their eyes, the lamp is no longer seen, yet they still know it is there, because the perceptual signal has returned into the internal model. That return is the loop closing. Without it, perception would scan forward forever, past the lamp, through the wall, and beyond, never resolving into knowledge. The loop closes exploration into understanding, and stabilizes reality within a field of perceptory confirmation.

Perceiver and perceived

Because the Boundary distinguishes interior from exterior, it distinguishes the perceiver from the perceived. Western philosophy has worked this distinction for centuries. Descartes made the perceiving subject foundational and the world that which it knows by inference; Kant made the structure of perception itself the condition of any object of experience; Husserl's phenomenology made consciousness always consciousness of something, so that the subject-object relation is constitutive of consciousness rather than something it encounters.

The framework's contribution is to treat the subject-object distinction not as an epistemological puzzle, and not merely as a feature of consciousness, but as the structural architecture of perception at every scale, the architecture the Boundary instantiates. The perceiver is what the closure produces from within; the perceived is what it produces as its outside; neither is prior, and the Boundary is what makes them what they are.

This generalizes beyond cognition. The boundary of a hadron distinguishes its three-quark closure from the surrounding field; a cell membrane distinguishes metabolic interior from environment; the boundary of a self distinguishes I from world; the boundary of a culture distinguishes us from them. Each is the same structural element operating at a different substrate.

At the cognitive scale the distinction takes a precise form, which George Herbert Mead captured as the I and the Me. The I is the active, perceiving self that meets the world; the Me is the perceived self, the internalized image built from one's relations with others. The two come into existence together, in the inseparable relation the Constant names, with the Boundary between them as the structural location at which self-perception closes its loop. The framework reads Mead's distinction as the Boundary operating at the cognitive scale, and reads the developmental psychology of self-formation in infants, the self established through differentiation from the caregiver, as the empirical signature of the same structural necessity.

Properties of the Inseparation Constant

The Constant is the most stable relationship among asymmetric Perceptory Fields, exceeded only by the perfect stability of the ground state itself. Because the inseparable pair comes into existence together and cannot be separated without dissolving the field that contains them, the relation is not subject to disturbance from within: the pair holds because the holding is what the pair is.

The Constant operates at instant. There is no sequence in which the perceived appears first and the not-perceived afterward; the two arise simultaneously with the closure of the field. This instantaneity follows from the framework's account of time, developed in Law X, where time is the perception of change within a Perceptory Network. The arising of the inseparable pair is not a change perceived; it is the structural condition under which any change could be perceived, and so is prior to time.

And the Constant constructs the frameworks within all perceived and not-perceived realities. Every Perceptory Field that exists carries its inseparable opposite at its boundary, so the Constant generates pairs at every scale of the hierarchy Law VI develops: particle and antiparticle at the matter scale, north and south at the magnetic scale, self and other at the cognitive scale, inside and outside at the social scale. It is the structural source of every observable inseparable pair the framework recognizes.

The Constant in observable phenomena

The claim that the Constant is universal can be tested against the inseparable pairings nature exhibits, each, on the framework's reading, the Constant operating at a particular substrate.

Magnetism is the most familiar. North and south poles do not exist independently; cut a bar magnet in half and the field regenerates new poles at the cut, because polarity belongs to the field, not the metal, and the field requires both poles for its closure. The case of magnetic monopoles is where the Constant's universality has been most strongly tested. Theory has long allowed monopoles, particles carrying a single magnetic charge with no opposite pole attached, yet decades of searching have not found them, with one possible and contested detection in 1982. The framework reads this record directly: the Constant rules out genuinely isolated monopoles. What appears as a local monopole must have its inseparable opposite somewhere within the larger field, perhaps at great or even cosmological distance, but present as a structural necessity. A truly isolated monopole would violate the Constant, and the framework predicts that none will ever be observed.

Particle and antiparticle pairing is the same rule at the matter scale. Every charged fermion has its antiparticle, generated at the boundary of the field that constitutes the matter scale, and the pair cannot exist independently because each is what the other is not. Cases where a neutral boson is its own antiparticle the framework reads not as genuine self-pairing, which its structure excludes (Appendix B), but as a pair whose members coincide in every quantum number while remaining conjugate modes of one field. Pair production and annihilation, in the reading Law VI develops, are the formation and collapse of the field that holds the pair, not the creation or destruction of two independent things.

At the most fundamental level the Constant joins existence to non-existence as the original inseparable pair. This pair arises not in the ground state, which has no asymmetric boundary and stands outside the Constant's domain, but at the boundary of every field that has departed from it: where a field closes asymmetrically, the perceived, existence, and the not-perceived, non-existence seen from within, come into being together. Existence and non-existence are not independent categories but the foundational inseparable pair the Constant names.

What the Constant rules out

The Constant has a sharp negative consequence: genuine isolation does not exist. No Perceptory Field with an asymmetric boundary can exist alone, because the boundary itself produces the inseparable opposite. The only exception is the Perceptory Field of Nothingness, which has no asymmetric boundary and so lies outside the Constant's domain. For all observable existence the Constant is universal, and it explains what would otherwise require independent justification: why observable existence always comes in pairs, why every fundamental force shows opposite-pole structure, why every self requires an other, why every particle has an antiparticle. Pairing is not an additional property of observable existence; pairing is what observable existence is, at its boundary, by structural necessity.

What the Boundary makes possible

The Boundary makes several things structurally possible. It makes identity possible: because the Boundary distinguishes a field from its surround, a field can be a distinct thing rather than collapsing into the symmetric ground state. It makes observable reality possible: because the feedback loop stabilizes perception into knowledge, the world can be registered as something that can be known, rather than dissipating unperceived. And it makes the registration of change possible: because the Boundary dynamically maintains the perceived and not-perceived distinction, asymmetries within the field can be registered as change, which is what Law X develops as time, the perception of change within a Perceptory Network.

A fourth possibility, interaction, depends on the Boundary being not merely dynamic but breachable, the condition under which one field comes into entangled relation with another. That breach is the hinge on which the next stage of the framework turns, and it is developed where it does its work: in the encounter of fields that produces new fields and mass, and in the formation of Perceptory Networks.

What Law IV has established

The Inseparation Constant is the structural rule that governs the boundary of every Perceptory Field that has departed from the symmetric ground state: wherever a field has an asymmetric boundary, the boundary produces an inseparable pair, perceived and not-perceived, that come into existence together and cannot be separated without dissolving the field. The Constant is the most stable relationship within asymmetric existence, operates at instant, and constructs the frameworks within all perceived and not-perceived realities. It is observable in every inseparable pairing in nature, from magnetic polarity through particle and antiparticle to the cognitive distinction of self and other, and it rules out genuine isolation; the sole exception is the Perceptory Field of Nothingness, which has no asymmetric boundary.

The Perceptory Boundary is the structural element at which the Constant operates: asymmetric, located at the closure-perimeter, and dynamic rather than static. As a self-regulating feedback loop it stabilizes perception into knowledge, and at the cognitive scale it takes the form Mead named the I and the Me. The Boundary makes identity, observable reality, and the registration of change possible; and through its breachability, developed where fields encounter one another, it makes interaction possible.

What remains is to develop what happens when two such fields, each with its own Boundary, encounter one another: the interaction their mutual breaching produces, the new field it forms, and the mass that field must spend energy to sustain. This is the work of Law V.

V Law of Consciousness

The interaction of Perceptory Forces forms a Perceptory Field, and its mass is the asymmetry the Forces carry into it.

Orientation

Law IV established the inseparable pair and the Perceptory Boundary at which it sits, and closed by naming the one capacity the Boundary holds that the rest of the framework depends on: that a Boundary can be breached, so that one field comes into entangled relation with another. Law V takes up that capacity and follows it to its most consequential result. When two fields that have departed from the symmetric ground state encounter one another, a new field is formed, and that new field must hold together the asymmetries it now contains. The energy it spends doing so is what the framework calls mass. This Law develops the encounter, the field it produces, and mass as the sum of stabilized asymmetries.

From one ground state to many distinct fields

Law III established that the ground state is one: across every direction infinity contains, each encapsulation of Nothingness is identical, and a plurality of identical, featureless closures is one closure met from infinitely many directions. Genuine plurality begins only when symmetry breaks. A break gives a field a feature, an asymmetry it did not before possess, and a field that perceives an asymmetry is no longer identical to the ground state or to any other field carrying a different asymmetry. After the break there are genuinely many fields, because at last they differ. What first breaks the symmetry the framework does not claim to derive; that trigger is the open question Law X records. But once there are distinct fields, the rest follows structurally, and it is the work of this Law.

The encounter, and why interaction requires Force

Two Perceptory Fields cannot act on one another directly. A field is a closure, and what reaches beyond a closure is not the field itself but its Perceptory Force, the outward-directed mode that arises when symmetry breaks (Law II). Two fields interact only when their Forces engage; without that engagement they remain separate closures, each sealed in its own self-perception. Interaction is therefore always an interaction of Forces, and it takes two forms, distinguished by what the Force meets: another Force, or an existing Field. The first form creates; the second modifies.

Force upon Force: the creation of a new field

When the Perceptory Force of one field meets the Perceptory Force of another, the meeting does not alter either field; it produces a new one. A closure forms around the encounter itself, a Perceptory Field that did not exist before. This new field differs from the Perceptory Field of Nothingness in the decisive respect: where the ground state perceives only Nothingness, and is therefore featureless and symmetric, the new field perceives asymmetries, the asymmetries the two Forces carried into the meeting. It is a closure, as every field is, but a closure around difference rather than around Nothingness.

Its creation and its mass are one event. The new field is born holding the asymmetry the two Forces brought together, and that held asymmetry is its mass from the first instant. There is no moment at which the field exists without mass and then acquires it; to be created from the meeting of Forces is already to hold their asymmetry, and to hold asymmetry is to have mass. The stabilizing geometry by which the closure holds, the third element that secures an inseparable pair into a standing field, is the rule of three of Law VI; here it is enough that the meeting of two Forces produces a closure that contains asymmetries and must hold them.

Force upon Field: the modification of an existing field

When a Perceptory Force meets not another Force but an existing Field, it does not create a new field; it disturbs and reconstructs the one it meets. The Force carries asymmetry, and in breaching the field's Boundary, penetrating it by perception without destroying it, for then the field would dissolve, it deposits that asymmetry into the field. The field is reconstructed: its internal asymmetry is no longer what it was. And because mass is the sum of the asymmetries a field must hold, the field's mass changes with its asymmetry. The more asymmetry a Force delivers into a field, the more that field must stabilize, and the greater its mass becomes. This holds at every scale. A particle, an atom, a larger body, each is a Perceptory Field whose mass rises as the asymmetry deposited into it rises.

When two fields each direct Force upon the other, the interaction becomes mutual, and the two fields are entangled: each has registered the other within its internal model, each Boundary breached by the other, so that neither can change without the change propagating to the other through the breach. This is what entanglement structurally is, in the framework's reading: not action at a distance, but the inseparability of Law IV now holding between two fields rather than within one.

The wave is carried asymmetry

What a Perceptory Force carries, in either form of interaction, is asymmetry, and this is why the propagating Force appears as a wave. A wave oscillates, rising and falling and rising again, and that oscillation is precisely the asymmetry the Force carries, set in motion. "Wave" is the name given to asymmetry travelling; the up-and-down is the carried asymmetry of the Perceptory Force made visible.

The clearest instance is the photon. A photon is a Perceptory Force in pure propagation, carrying asymmetry through space, and it has no rest mass because it never closes into a standing field of its own, carrying asymmetry without ever holding it in a closure. Only when a photon's asymmetry is absorbed by a field that does close, a Force-upon-Field interaction, does that asymmetry become mass, and the mass it becomes is the mass of the absorbing field, whose asymmetry has just increased.

Mass as the sum of stabilized asymmetries

A field that perceives asymmetries must hold them together. An asymmetry is a departure from the balance the ground state keeps for nothing, and to persist as a single closure rather than come apart, a field must continually spend perception, and so energy, on stabilizing the asymmetries it contains. This expenditure is what the framework names mass. Mass is not a substance a field is made of; it is the energy a field spends holding its own asymmetries in a stable closure, the sum of the asymmetries it must perceive without pause in order to remain one thing. A field comes by its asymmetries in the two ways the interactions describe: it is born holding them, when it is created from the meeting of two Forces, and it gains further asymmetry, when a Force deposits it through the Boundary. Either way, its mass is the sum of what it must hold.

This reading stands closer to what physics has found than the older picture of mass as quantity of matter. The mass of a proton is not, in the main, the mass of the quarks inside it. The rest masses of its three quarks, the portion attributable to the Higgs mechanism, come to roughly one percent of the proton's mass; the other ninety-nine percent is the energy of binding those quarks together, the energy of the gluon field that confines them and the motion of the quarks within, as lattice-QCD calculations of the proton's mass have made precise. Almost the whole of the proton's mass is the energy its internal structure costs to maintain. The framework reads this as its own claim seen in the laboratory: a proton's mass is the energy it spends stabilizing the asymmetries bound within it, the sum of those asymmetries carried as energy, mass and energy being one quantity in different dress. Mass is what it costs a field to keep its asymmetries closed into one existent.

The sharpest confirmation that mass is stored asymmetry comes from the attempt to isolate a quark. A quark cannot be pulled free of the hadron that holds it. As it is drawn away, the colour field between it and its neighbours stretches into a narrow tube whose energy rises in proportion to the separation, and at a certain distance it becomes cheaper for the field to make new matter than to stretch further: the tube breaks, and a new quark and antiquark appear at the broken ends. The stored energy of the stretched field is converted directly into the mass of the new quarks, so that the attempt to expose a single quark yields not a lone quark but new hadrons. The framework reads this as its claim shown at its sharpest, stored field energy seen turning into the mass of particles that did not exist before. And the impossibility of isolating a quark is the impossibility of exposing a bare, unbalanced asymmetry: the field answers every such attempt by producing the partner that restores its balance, never a lone unbalanced charge.

Asymmetry, stability, and decay

If mass is the energy a field spends stabilizing the asymmetries it holds, then asymmetry and stability stand in a definite relation: the more asymmetry a field must hold, the more there is to stabilize, and the more readily it comes apart. The perfectly symmetric ground state, holding nothing, is perfectly stable, as Law III established, and stability falls as asymmetry rises. Decay, on this reading, is a field shedding asymmetry to reach a more stable, lower-asymmetry configuration, the structural counterpart of a strained thing relaxing toward rest.

Among elementary particles the pattern is clean, and it is what physics observes. Heavier particles are unstable and decay into lighter ones, while the lightest particles, having nothing lighter to decay into, are stable: the electron and the proton endure, where the heavier muon, the tau, and the W and Z carriers decay rapidly. Physics describes decay precisely as a relaxation toward lower mass, an unstable particle transforming into lighter and more stable products. The framework reads this directly: a heavier particle holds more asymmetry, and sheds it by decaying toward configurations that hold less.

The connection runs deeper than the rough correlation of mass with instability, into the relation between symmetry and stability that physics already recognizes. The stable particles are stable because each is protected by an unbroken symmetry: the electron's stability follows from the conservation of electric charge, the proton's from the conservation of baryon number, and in every known case a vanishing tendency to decay is traced to a conservation law that an unbroken symmetry enforces, with instability following where that symmetry is broken. This is the framework's own claim in the vocabulary of the Standard Model: symmetry is stability, and broken symmetry, which is to say asymmetry, is what decays.

One caution keeps the claim exact, and it concerns composite nuclei rather than elementary particles. It is not true that the heaviest nucleus is always the least stable, nor the lightest the most stable. The binding energy per nucleon, the measure of how tightly a nucleus is held, rises with mass to a peak around iron and nickel and falls away on either side, so that the most stable nuclei are of middling mass: lighter nuclei gain stability by fusing toward iron, heavier nuclei by fissioning toward it. The framework reads this without strain, because for a composite the governing quantity is not total mass but how efficiently the field binds the asymmetries it holds. Iron is the most efficient such closure, the size at which the short-range binding reaches across the whole nucleus; beyond it the asymmetry outruns what the binding can hold, and stability declines. The simple form of the principle, more asymmetry and less stability, holds cleanly for elementary particles; for composites it takes the form that stability tracks the efficiency with which asymmetry is bound, which is greatest at iron.

Cautions

Two cautions hold the claim to the evidence. The framework reads the dominant, dynamical part of mass, the binding energy that is most of every hadron, as the energy of stabilizing asymmetries; it does not contest the small remainder, the quarks' rest mass, which arises as physics says from the Higgs mechanism, itself a symmetry breaking in the framework's reading, so that even that remainder is asymmetry expressed as mass. And the full development of mass, its relation to gravity and to the curvature of the perceptory field, belongs to Law VIII. Law V establishes the principle, that mass is the energy of stabilized asymmetry, and the two interactions that yield it: Force meeting Force, which creates a field already holding its mass, and Force meeting Field, which raises a field's mass by the asymmetry it deposits.

What Law V has established

Two Perceptory Fields cannot act on one another directly, only through their Perceptory Forces, and the interaction of Forces takes two forms. When Force meets Force, a new Perceptory Field is created, a closure around the encounter that perceives asymmetries rather than Nothingness; its creation and its mass are one event, since it is born holding the asymmetry the two Forces carried in. When Force meets an existing Field, no new field is made; the Force deposits the asymmetry it carries through the field's Boundary, reconstructing the field and raising its mass, the more asymmetry delivered, the greater the mass, at the scale of particles, atoms, and bodies alike. What a Force carries is asymmetry, which is why it propagates as a wave, the oscillation being that asymmetry in motion; the photon carries asymmetry without ever closing into a standing field, and so has no rest mass until its asymmetry is absorbed by a field that does close. Mass, in either case, is the energy a field spends stabilizing the asymmetries it holds, a reading borne out by the proton, almost all of whose mass is binding energy rather than the rest mass of its quarks. The stabilizing geometry of the closure is the rule of three, the work of Law VI; the relation of mass to gravity is the work of Law VIII.

What remains is to develop how the encounter's closure is held together, how entanglement forces hierarchy, and how the rule of three emerges from three-point closure. This is the work of Law VI.

VI Law of Consciousness

Every stable existent is a closure of three, two elements held inseparable by a third; from such closures, entanglement and containment weave every Perceptory Network.

Orientation

The preceding Laws have established the primitives. Law I demonstrated that for Nothingness to exist it must be perceived. Law II identified the Perceptory Force as the fundamental act through which existence is registered. Law III established that the Perception of Nothingness perceives itself, generating the first Perceptory Field. Law IV showed that everything that exists is constructed through the inseparable interaction of perceived and not-perceived, the Inseparation Constant, residing at the Perceptory Boundary of every departed Perceptory Field, and developed the Boundary itself as a dynamic feedback loop that stabilizes perception into knowledge. Law V demonstrated that fields interact only through their Perceptory Forces, and that this interaction takes two forms: Force upon Force, which creates a new field born already holding the asymmetry the two Forces carried into the meeting, its creation and its mass a single event; and Force upon Field, in which a Force deposits the asymmetry it carries through the field's Boundary, reconstructing the field and raising its mass. Mass, in either case, is the energy a field spends stabilizing the asymmetries it holds.

Law VI now poses the question that follows from these foundations: once fields exist and can interact, how do they organize? Two fields need not interact at all. They can coexist, each sealed in its own self-perception, holding different asymmetries and never touching, until a Force is directed from one to the other. When they do become bound, the framework finds that they are bound in just two ways. They are entangled, fused as peers into a single unit, or one is contained within the other, nested as a part within a whole. These two relations, entanglement and containment, are the whole vocabulary of perceptory organization, and from their weaving every network, every hierarchy, every structure the universe exhibits is built. The central finding of this Law is that the two are not separate primitives. Containment is entanglement that has become asymmetric, so that hierarchy is not imposed upon entanglement from outside but is what entanglement becomes when direction enters it.

The two relations: entanglement and containment

Two Perceptory Fields are entangled when a change in one is bound to a change in the other, not by a signal passing between them but by mutual registration, each field holding the other within its own internal model. This is the structural correlation Law V established: once two boundaries have been breached into one another, neither field can change without the change propagating to the other through the breach. The empirical phenomenon this aligns with is quantum entanglement, which permits correlation between distant measurements without permitting communication between them, an alignment the framework's reading preserves rather than violates.

Entanglement, in its simplest form, is symmetric. When two fields are entangled as equals, neither contains the other; they are fused side by side into a single unit. This is the relation at work when two concepts become one in thought. A cup used always for coffee becomes, not a cup described as holding coffee, but a coffee cup, a single unit in which the two concepts are so bound that neither is fully thought without the other. To think bedroom is to summon bed and lamp and the rest, already entangled into one. The unit is more than its parts set side by side; it is the parts fused into something that can itself be perceived as one. Symmetric entanglement is how the perceptory world builds units.

Containment is what entanglement becomes when it ceases to be symmetric. When one of two entangled fields holds the other within it, the relation acquires a direction it did not have before: there is now an inside and an outside, a field that contains and a field that is contained. The quark is within the proton; the proton is not within the quark. This asymmetry is the whole of what distinguishes containment from entanglement. Containment is still held together by entanglement, the contained field cannot change without the containing field registering it, exactly as Law V requires, but it is entanglement with a direction, and that direction is a level. Strip the asymmetry away and containment relaxes back into peer entanglement; add it, and a level appears. Hierarchy is nothing other than asymmetric entanglement, containment stacked upon containment. Here, once again, asymmetry is the thing that builds: symmetric coupling lies flat, and only when direction enters does structure rise into levels.

The clearest display of stacked containment is the cascade that runs through every ordinary object. A cell contains its proteins; each protein contains its atoms; each atom contains its protons and neutrons; each proton contains its quarks. At every step a field is held within a larger field, and the whole descent is one chain of containment, each level entangled with the level that contains it. This is what a Perceptory Field of any complexity is, seen from within: not a single thing but a nested depth of contained fields, levels within levels, held at each step by the asymmetric entanglement that makes one field the inside of another.

A real perceptory network runs both relations at once. Across it lie lateral threads, fields entangled as peers into units; through it run vertical lines, units contained within larger units. From these two relations alone, woven in every combination, the endless diversity of perceptory structures is generated. And the network cannot stay flat. A configuration in which every field were merely entangled as an equal, none containing another, would hold no levels and no inside; it would be perfectly symmetric, and by the same principle that governs every other Law, unstable, for the moment one field registers another field's registering of a third, asymmetry has entered, and with asymmetry, direction, and with direction, containment. Flatness collapses into hierarchy because asymmetry is unavoidable wherever perception compounds.

Two consequences follow, one deferred and one immediate. The deferred one, developed in Law IX, is that multidimensionality is what this nesting looks like from outside: each level of containment is a perceptory dimension. The immediate one is that containment carries change in both directions. If a containing field changes, the change reaches the fields it contains, because they are entangled with it. This is why a change at the top of a chain reaches all the way to the bottom, why a perception can reconfigure the chemistry far beneath it. The body's cells perform their own perception and function on their own, yet a thought, belonging to the field that contains them, can reach down and alter their working, because to contain is to be entangled with, and entanglement carries change both ways. Downward causation is not a force reaching across a gap; it is a container reconfiguring what it holds.

The minimum stable field: the rule of three

The two relations describe how fields stand to one another. A further question concerns the single field itself: what does it take for one Perceptory Field to hold together as a closed unit at all, whether it is a peer in an entanglement or a level in a containment? The answer is a geometry, and it is not arbitrary; it is fixed by the minimum conditions under which a closure can sustain itself.

Two perceptory points, entangled, draw a single line between them. This is one-dimensional and flat. It cannot enclose anything. It cannot constitute an inside. It can only oscillate between its two endpoints, and any disturbance dissolves it.

Three perceptory points, entangled, draw three lines that meet at a centre. The three lines together describe the minimum closed figure: a triangle whose interior exists by virtue of the three boundaries that define it. Three is the minimum number of entanglement points that produces a stable closed Perceptory

Field. This is not metaphor. It is the geometric consequence of what entanglement is.

The rule of three is therefore not a feature of certain systems within reality. It is a structural condition for any stable existence whatsoever. Wherever a Perceptory Field persists, three-point hierarchical entanglement is operative within it.

Hadrons as the physical signature of the rule of three

The clearest physical confirmation of the rule of three appears at the matter scale. Protons and neutrons, the constituents of every atomic nucleus, and therefore of all ordinary matter, are bound states of three quarks. This is the simplest stable Perceptory Field that exists at the matter scale: three perceptory concentrations entangled at a single centre, forming a closed field that contains them.

The conventional account treats the three-quark structure of hadrons as an empirical fact described by quantum chromodynamics, with the colour-charge confinement principle explaining why isolated quarks are not observed. Within the framework of the Laws of Consciousness, the three-quark structure is not an empirical accident requiring an additional theoretical construct. It is the direct manifestation of the rule of three. Quarks are not particles arranged in a sphere; the quarks are the three-point hierarchical entanglement that makes the hadron a stable existent.

Higher numbers of perceptory concentrations are geometrically possible. Higher-order closed fields, four-point, five-point and beyond, can in principle form. They are observed to be short-lived, decaying toward three-point configurations, a pattern the framework reads as consistent with three-point minimality while acknowledging that a formal stability ordering across configurations remains to be derived (Appendix B). This is why three-quark configurations dominate in ordinary matter. The exotic multi-quark states observed in high-energy physics, tetraquarks, pentaquarks, are precisely the transient higher-order configurations the framework expects.

Inseparable pairs and the third element

Law IV established that every Perceptory Field exists alongside its inseparable opposite, and that the Inseparation Constant resides at the boundary between them. The rule of three now permits a sharper formulation of what the Inseparation Constant is.

Two inseparable opposites, perceived and not-perceived, particle and antiparticle, north pole and south pole, self and other, cannot constitute a stable Perceptory Field on their own. Two points draw only a line, and a line cannot close a field. The inseparability of the pair therefore requires a third element: a perceptory entity that holds the pair in their inseparable relation. This third element is what the Inseparation Constant becomes when treated structurally rather than abstractly. It is the boundary made present. It is the mediating element through which the three-point closure holds.

I will name this third element the Inseparation Element. Every stable inseparable pair, at every scale of hierarchical entanglement, contains an Inseparation Element that constitutes its inseparation.

The Inseparation Element of magnetism is the field between the poles. The poles do not hold each other in opposition by themselves, they are held by the magnetic field, which is what makes the polarity inseparable. Cut a bar magnet in half and the field regenerates with new poles, because the Inseparation Element is what produces the polarity, not the reverse.

The Inseparation Element of self and other is the act of perception that joins them. Self and other are not two independent entities that subsequently engage in perception; they are two perceptory points held in inseparable relation by the Perceptory Force itself, which constitutes their boundary.

The Inseparation Element of mother and child is the bond, and bond is not a metaphorical term here. It is a real perceptory entity at the biological scale. Severance of the bond does not produce two independent fields; it produces a wound, because the Inseparation Element has been disrupted, not merely the relation it supported.

The A-particle: a committed prediction

At the scale of particle physics, the framework predicts a specific instance of the Inseparation Element. For every particle-antiparticle pair, there exists a third element that constitutes the pair's inseparation. I name this the A-particle.

The A-particle is not a fourth class of entity in addition to particles and antiparticles. It is the Inseparation Element generated at the centre of the particle-antiparticle relation, the structural object at which the three lines of the minimum stable Perceptory Field meet. Mainstream physics has not catalogued the A-particle because mainstream physics treats particles and antiparticles as the only two elements requiring explanation, and treats their relation as a property rather than an entity.

The framework predicts that pair-production and pair-annihilation events should exhibit distributional anomalies relative to the complete predictions of standard quantum electrodynamics, deviations in spectral shapes, angular correlations, branching ratios, or timing at fixed total energy, deviations that correspond, in the framework's reading, to the formation or collapse of the A-particle as the three-point Perceptory Field reorganizes. Annihilation, in this account, is not the simple mutual cancellation of particle and antiparticle. It is the collapse of the three-point closed structure that the A-particle holds together, with the distribution of the collapse's fixed energy budget across observables carrying the signature of the A-particle's role as Inseparation Element.

This is a falsifiable prediction in the sense that pair-annihilation energetics are measurable and the framework's prediction is that they should exhibit anomalies beyond what standard theory predicts. If high-precision pair-annihilation experiments confirm that the energy budget of pair-annihilation agrees, across the full set of measurable observables, with the complete predictions of quantum electrodynamics at the highest precision attainable, the prediction is progressively constrained; the framework acknowledges that, until it supplies a predicted magnitude for the anomaly, null results bound the prediction without strictly falsifying it, and that its scientific weight remains limited until that quantitative work is done. If anomalies are detected, the framework's reading of pair-annihilation as three-point Perceptory Field collapse becomes a candidate for further investigation.

A specific clarification deserves emphasis here. The A-particle, in the framework's reading, is not a free-state entity that could be observed independently of the particle-antiparticle pair it holds together. As the Inseparation Element of an inseparable pair, the A-particle exists only within the closed three-point structure of the Perceptory Field that contains it. Any registration event that would measure the A-particle as a separate object would, by Law VI's own architecture, introduce a fourth perceptory entanglement into the three-point structure and reconfigure it, collapsing the very Inseparation Element that registration would attempt to detect. The A-particle is therefore constitutively unmeasurable as a free-state particle. What is measurable is its collapse signature: the deviation, within a fixed total energy budget, in the distribution of pair-annihilation observables that the three-point-to-two-point reconfiguration produces. This places the A-particle in a structural position analogous to virtual mediators in quantum field theory, entities that are not directly observable as free states but whose effects on measurable quantities (scattering amplitudes, in QFT's case; pair-annihilation observables, in the framework's case) can be characterized.

What the framework does not yet specify, and what cannot be specified by direct measurement of the A-particle as a free state: its mass, spin, electric charge, gauge coupling, lifetime, decay channels, and conservation-law behaviour. These quantities, in the framework's reading, can be characterized only theoretically, through the contribution they make to pair-annihilation observables under specific theoretical decompositions of the energy budget. Future formal-mathematical development must identify the structure under which such contributions can be extracted from the measurable collapse signature; this is the framework's call on theoretical physics with respect to the A-particle prediction.

Wave-function collapse

Law II read the Observer Effect as registration producing a definite field rather than disclosing a pre-existing particle, and Law V gave the wave its nature: the Perceptory Force in propagation, carrying its asymmetry, not yet closed into a standing field. The two readings meet here. The wave and the particle are not two things but the two modes of one Force, the inseparable pair of Law IV at this scale: the not-perceived, the Force still propagating and unregistered, and the perceived, the closed field a registration produces. Before observation the Force propagates with the particle latent within it, the perceived held inside the not-perceived, in the three-point closure the rule of three requires, the Inseparation Element holding the pair in superposition. This is what superposition is, not a paradox but the inseparable pair before registration has drawn one side of it into actuality.

Observation is the introduction of a new entanglement. When the observer's Perceptory Field registers the system, the Force closes, the propagating wave registered into a definite field, the perceived actualized out of the not-perceived. Said in the vocabulary of the closure, a fourth element has entered a three-point structure and displaced the Inseparation Element that held wave and particle in superposition, so that the system re-closes, now bound to the observer rather than to the original pair. These are one event in two languages: registration closing the Force into a field, and a fourth entanglement reconfiguring a three-point closure. Collapse is therefore neither mysterious nor a making of something from nothing; it is the closure of a Force that was already carrying its asymmetry, the perceived drawn out of the not-perceived by the act that registers it.

The hierarchy chain across scales

The rule of three operates recursively. Each stable level is a three-point closure, and each closed field then becomes a contained field within the larger field at the next scale, so that the chain is one of containment stacked upon containment, the relation this Law has placed at the root of hierarchy. The visible record of this recursion traverses every order of magnitude in the known universe.

Three quarks form a hadron. Hadrons and electrons form atoms, and here too the framework expects three-point structure: the nucleus and the electron field as an inseparable pair, held by a third structural element, the Inseparation Element of their binding, whose physical identification remains open. Atoms form molecules. Molecules form proteins, the smallest level at which biological perceptory architecture emerges. Proteins form cells. Cells form organs. Organs form organisms. Organisms form minds. Minds form cultures. Each transition runs both relations at once. The constituents entangle with one another as peers and fuse into a single unit, and that unit stands as a new level that contains them, the lateral entanglement making the many into one and the containment making the one a level above the many. Two hydrogen atoms and an oxygen atom, entangled into the single unit we call a water molecule, are then contained within it as surely as the molecule is contained within the cell that holds it, the same doubled movement repeating at every step of the chain.

The chain is not arbitrary. It is the universe's record of the two relations operating across scales, each level's constituents entangled into a unit and each unit contained within the next, every level's stability earned by satisfying the three-point closure the rule of three requires. What appears as the diversity of observable reality, particles, atoms, organisms, civilizations, is a single architecture, two relations and one closure geometry, repeating at successively larger scales of perceptory organization.

Hierarchy in biology: DNA and memory

The rule of three, and the two relations that organize it, are not confined to physics. At biological scale they produce phenomena that mainstream biology has documented but not yet recognized as instances of the same structural principle.

Consider DNA as a dynamic Perceptory Field rather than a static blueprint. DNA does not merely encode genetic information; it perceives chemical signals within its environment, and gene expression is modulated accordingly. Environmental factors, diet, stress, exposure to specific compounds, alter gene expression without altering the underlying DNA sequence. This is the field of epigenetics, and it is the standard account. The framework reads it as containment running downward: the DNA field is contained within the cell, the cell within the organism, the organism within its social and ecological surround, and because to contain is to be entangled with, a change in any containing field reaches the fields it holds. Gene expression is the visible signature of those nested containments reaching down to the molecular scale. Mutation under environmental stimulus, radiation or chemical exposure, is the framework's downward causation at its sharpest: the Boundary of the DNA field reconfigured by a new entanglement arriving from the field that contains it.

Human memory exhibits the same architecture. Each memory is stored through perceptory entanglements in the brain, and exists as a tangible aspect of reality only when perceived or recalled. Therapeutic methods that access traumatic memory, EMDR, Brainspotting, somatic approaches, operate by deliberately re-engaging the perceptory fields in which the memories are held, permitting their reorganization.

The clinical phenomenon of dissociative identity disorder makes the architecture especially visible. In dissociative configurations, shifts between identity states can be accompanied by measurable changes in posture, voice, autonomic regulation, pain response, allergic presentation, memory access, and somatic self-representation. These are downward causation made clinically visible: the dominant cognitive Perceptory Field has reconfigured, and the lower-level fields it contains, no longer held in their prior configuration, express themselves directly. They are not produced by conscious choice; they follow from the containing field's reach into what it holds. The hierarchy has not collapsed; it has reorganized. What presents at the biological scale is the visible signature of a three-point closure rearranging its centres.

These cases are not exotic curiosities. They are the rule of three, and the two relations that organize it, operating at biological scale, with the same structural logic that governs hadrons.

What Law VI has established

Perceptory fields are organized in two ways, and only two: by entanglement, in which fields are fused as peers into a single unit, and by containment, in which one field is held within another. Containment is entanglement that has gone asymmetric, so that hierarchy is not a separate principle but the directed form of entanglement itself, and a perceptory network is the weaving of these two relations, lateral and vertical, across every scale. The minimum stable field, whether it stands as a peer or as a level, is a three-point closure, with two inseparable elements held in their inseparation by a third, the Inseparation Element. At the scale of particle physics, the framework predicts the A-particle, with the further prediction that pair-annihilation events exhibit a corresponding energy signature. Wave-function collapse is the reconfiguration of an Inseparation Element when a fourth perceptory entanglement is introduced into a three-point structure. The chain of nested fields from quark to culture is the universe's record of containment operating recursively across scales, and the reach of a containing field down into what it contains is the structural ground of downward causation.

What remains is to ask what the totality of all hierarchical entanglements amounts to. If every stable existence is a three-point closed Perceptory Field, what geometric form does the totality of all such fields together constitute? This question is developed across the remaining Laws: Law VII first addresses the force-structure produced by three-point closure (antipodal pairing); Law X develops the largest-scale geometry as the totality of perceptual infinity, with the Perceptory Sphere as its symbolic representation; and Appendix B presents the equations that capture the Sphere's structural identities.

VII Law of Consciousness

Every force is an antipodal pair held by its Inseparation Element.

Orientation

Law VI established how Perceptory Fields are organized: by two relations, entanglement and containment, the second being entanglement that has gone asymmetric. It also established the geometry of the single stable field, the three-point closure, two inseparable elements held in their inseparable relation by a third, the Inseparation Element. This closure geometry was derived structurally; it was confirmed at the matter scale by the three-quark composition of hadrons; it was generalized across scales as the rule of three.

Law VII now poses the question that follows directly. If three-point closure is the geometry of stable existence, what is the force-structure this geometry produces? When perceptory entanglement operates between the points of a stable Perceptory Field, what does it do, and what does that doing look like from outside?

The answer follows necessarily from the geometry itself: antipodal pairing. Every force expressed at one perceptory point manifests its complementary expression at the antipodal point, and the two expressions together constitute a single antipodal-pair force.

From three-point closure to antipodal arrangement

The derivation is short and structurally tight. From Law VI, a stable Perceptory Field consists of three perceptory points entangled at a centre. Two of those points are the inseparable opposites that Law IV identified, perceived and not-perceived, or any of the inseparable pairs the framework has now named (particle/antiparticle, north/south, self/other). The third point is the Inseparation Element, which holds the pair in their inseparable relation.

The geometry has only one configuration that produces a stable closed field with this structure. The two inseparable elements must sit at antipodal positions, diametrically opposite each other within the field, with the Inseparation Element at the centre between them. Any other arrangement either fails to close the field or fails to maintain inseparability.

Two points placed adjacent rather than opposite cannot constitute an inseparable pair, because adjacency permits drift toward fusion or separation. Antipodal placement is the only configuration in which two elements remain maximally distinct yet are held in stable relation through a mediating third element balanced exactly between them. The Inseparation Element is what enables both maximum distinction and maximum stability, it is structurally balanced between the two poles, leaning toward neither (Appendix B's balance condition), and its presence is what prevents the pair from collapsing into one another or dispersing.

This is not a metaphysical claim. It is a geometric one. If the rule of three from Law VI holds, then the inseparable opposites of Law IV are necessarily antipodal. Opposite in your framework means literally that, across from, on the far side of the Inseparation Element, paired through the geometry of the closed perceptory field.

Force as entanglement-in-action

A definition that has been carried implicitly through the preceding Laws now requires explicit statement.

A force is what perceptory entanglement does between points. This is the load-bearing definition the rest of the chapter rests on. Forces are not separate ontological entities additional to perception, fields, and entanglement. Forces are not properties that exist alongside Perceptory Fields. Forces are the visible expression of perceptory entanglement when that entanglement operates between points within a stable Perceptory Field.

Mainstream physics treats forces as fundamental ontological entities, the four fundamental forces are taken as basic features of reality, alongside particles and fields. The framework of the Laws of Consciousness inverts this. Perception is fundamental. Entanglement is what perception does when it perceives itself in relation. Force is what that entanglement looks like from outside, when its action between points becomes observable.

This collapses a distinction that has structured Western physics since Newton: the distinction between what is and what acts. In the framework, what is and what acts are the same thing seen from different angles. Existence is entanglement; entanglement is action; action observed is force. There is no separate stratum of forces operating on a substrate of matter. The substrate and the forces are the same perceptory phenomenon, viewed from inside (as existence) and from outside (as force).

The Law of Antipodal Force

The two derivations now combine. If forces are what entanglement does between points, and the geometry of three-point closure forces those points into antipodal arrangement, then forces themselves must exhibit antipodal-pair structure.

Stated as the Law: every fundamental force operating through finite-dimensional Perceptory Force modes exhibits antipodal-pair structure. The pair is not two independent forces. It is a single force with two complementary expressions, held in inseparable relation by an Inseparation Element at the centre between them.

The framework therefore predicts: no fundamental force operating through a finite-dimensional Perceptory Force mode exists without an antipodal partner. Discovery of such a force would falsify the framework at this point. Conversely, every observed force should reveal antipodal-pair structure when examined with sufficient precision.

This prediction is not exotic. Mainstream physics has already documented antipodal-pair structure in every fundamental force it has studied. What the framework contributes is the recognition that these documented pairings are not separate features of separate forces. They are instances of a single structural rule operating at every finite-dimensional level. Physics has been describing the visible signature of a deeper geometry it has not yet named.

Confirmation in observable physics

The known fundamental forces that operate through finite-dimensional Perceptory Force modes each exhibit antipodal-pair structure in precisely the form the framework predicts.

Electromagnetism. Positive and negative charge are antipodal. The electromagnetic field generated between them is the Inseparation Element of the pair. What appears to be a single force with dual signs is, in the framework, a single antipodal-pair force whose two complementary expressions are mathematically related by sign reversal. The field between the charges is not incidental to the polarity, it is what produces and sustains the polarity. Without the Inseparation Element, charges cannot interact; they are merely two unrelated phenomena.

Magnetism. North and south poles are antipodal. The magnetic field between them is the Inseparation Element. Law VI already established this case in detail: cut a bar magnet in half and the field regenerates with new poles, because the Inseparation Element is what produces the polarity, not the reverse. A truly isolated magnetic monopole would require severing the Inseparation Element from the pair, a configuration the geometry resists, except potentially at extreme energy regimes where the entire three-point closure may reconfigure.

Strong force. The strong nuclear force exhibits a structure mainstream physics has documented as anomalous and the framework now reads as structurally consistent with the rule of three. Colour charge is three-fold rather than two-fold. Quarks carry one of three colour charges (red, blue, green); antiquarks carry the corresponding anti-colours. Baryons are colour-neutral combinations of three quarks. Mesons, bound quark-antiquark pairs, are the two-point-plus-Element case: their three-point closure runs through the gluon field as Inseparation Element, and their characteristic instability relative to baryons is what the framework's geometry expects of a closure carried by the Element alone.

The framework reads this exactly as it appears: colour confinement is the rule of three operating directly at the strong-force scale. The strong force is not like antipodal pairing with three poles instead of two. The strong force is the rule of three from Law VI made visible at the matter scale, with each colour antipodally paired with its anti-colour, and the three-fold structure constituting a complete closed three-point field. Quark confinement, the principle that isolated quarks have never been observed, is not a separate empirical mystery. It is the geometric necessity of three-point closure: a single quark cannot exist alone because a single point cannot constitute a stable Perceptory Field. The framework predicts confinement structurally rather than treating it as an additional axiom.

A specific objection deserves direct response here. A physicist may say the framework is mapping the number three onto QCD after the fact, that the three-quark structure of hadrons is fully accounted for by SU(3) gauge symmetry, and that the Rule of Three adds nothing to the existing theoretical apparatus. The framework's reply: it is not replacing the SU(3) account or the mathematics of QCD, both of which the framework leaves intact. It is proposing a deeper structural reason why the symmetry takes the form SU(3) takes, why three-fold rather than four-fold or five-fold gauge symmetry is what becomes stable at the matter substrate. The Rule of Three predicts that wherever stable closure is observed, the underlying structural geometry will exhibit three-point closure or its compound forms; SU(3) at the matter scale is one specific instantiation of that structural commitment. The empirical persistence of three-quark configurations in stable matter, and the empirical transience of four-quark and five-quark configurations, is what the framework predicts to be observed and is what the experimental record shows.

Weak force. The weak interaction exhibits antipodal-pair structure with one critical difference from the other forces: massive asymmetry between the two antipodes. Left-handed and right-handed chirality are the antipodal pair, but in the universe as currently observed, the weak force interacts almost exclusively with left-handed particles. The right-handed antipode is suppressed.

This is one of the deepest puzzles of the Standard Model, known as parity violation, it has resisted clean theoretical explanation for seventy years. The framework reads it as follows: both antipodes exist; the antipodal-pair rule is not violated. What is observed is amplitude asymmetry, not antipodal absence. The right-handed weak interaction is suppressed under current low-energy conditions, but the framework predicts that at sufficiently high energy regimes, energies approaching unification scales or those obtaining in the very early universe, parity should be restored. This places the framework alongside left-right symmetric extensions of the Standard Model, proposals such as the Pati-Salam model, SO(10) grand unification, and other left-right symmetric models that already predict parity restoration at high energies. The framework does not propose a specific formal extension; it predicts that some such extension will be empirically supported, with parity restoration emerging at the energy scale at which the underlying symmetry is recovered, with left- and right-handed weak interactions exhibiting equal amplitudes. This is testable in principle through future high-energy collider experiments or through precise observation of conditions in early-universe cosmological signatures. The structural mechanism behind the restoration is developed in Law X: parity is recovered when the restoring coupling among the totality's latent triads becomes dominant at high energy, so that the prediction follows from the dynamics of breaking and restoration rather than standing as a separate stipulation.

This prediction is sharp and falsifiable. If parity remains permanently broken at all accessible energy scales, with no convergence toward restoration as energy increases, the framework's reading of parity violation is wrong and must be revised at this point.

Particle and antiparticle pairing. The general rule is that every charged fermion has a corresponding antiparticle, with some neutral bosons their own antiparticles. This is the most fundamental antipodal pairing in observable physics, and the framework now recognizes it as the direct geometric expression of the Law of Antipodal Force at the matter scale. The Inseparation Element of every particle/antiparticle pair is the A-particle named in Law VI, the third element that constitutes the pair's inseparation. Pair production and annihilation events are not the creation and destruction of two independent entities; they are the formation and collapse of three-point Perceptory Fields whose Inseparation Elements emerge or disperse along with the pair.

What Law VII has established

Forces are what perceptory entanglement does between points. The geometry of three-point closure, established in Law VI, forces inseparable pairs into antipodal arrangement with an Inseparation Element at the centre. Therefore every fundamental force operating through finite-dimensional Perceptory Force modes exhibits antipodal-pair structure. This structure is observable across electromagnetism, magnetism, the strong and weak nuclear forces, and the matter/antimatter divide. The framework predicts that no fundamental finite-dimensional force exists without antipodal partnership, and that the apparent parity violation of the weak interaction is amplitude asymmetry under current conditions, with parity restoration predicted at sufficiently high energy regimes.

What remains is gravity, the one fundamental interaction the framework does not treat as a finite-dimensional force mode. It is the subject of Law VIII.

VIII Law of Consciousness

Gravity is the act of perceptory entanglement between networks of the same structural kind.

Orientation

Law VII established that every fundamental force operating through finite-dimensional Perceptory Force modes exhibits antipodal-pair structure, and set gravity aside as the one fundamental interaction that is not such a mode. Law VIII now takes gravity up directly.

Gravity is the one force with no opposite charge. The others pair across a sign: positive and negative, north and south, colour and anti-colour. Gravity has no negative counterpart, no anti-gravity to set against its attraction, a point recently put to the test when the ALPHA experiment at CERN released antihydrogen in free fall and watched it fall downward, like ordinary matter, rather than rise. The accelerating expansion of the universe is a repulsion of another kind, the gravitational effect of the energy of empty space, not a gravitational charge of opposite sign. The framework reads this absence not as a failure of antipodal structure but as a sign that gravity's antipodal pair lies deeper than any difference of sign. The pair is not attraction against repulsion. It is entangling against disentangling, the dynamics not of paired points within a single field, from which Law VII exempted gravity, but of whole Perceptory Networks, which this Law now develops.

What gravity does

Gravity is the act of perceptory entanglement between Perceptory Networks.

When two networks of significant entanglement count come into proximity within the same substrate, they entangle. The visible signature of that entanglement, at the matter scale, is what mainstream physics has named gravitational attraction. But gravity is not pull, in the mechanical sense. Pull is what entanglement looks like from outside. From inside the framework, gravity is the event of one network registering another network of comparable structural kind, and the resulting entanglement that follows.

Seen from within, what reaches outward is the Perceptory Force itself, directed beyond a network's boundary by the asymmetries the network holds. A network laden with asymmetry, which is to say with mass, reaches outward and registers the networks around it, and that reaching, viewed from outside, is the pull we call gravity. The force reaching out and the entanglement drawing in are not two processes but one process seen from two sides.

This closes a thread Law V left open. Mass, Law V established, is the energy a field spends stabilizing the asymmetries bound within it, and those stabilized asymmetries are the network's internal entanglements, their total being the entanglement count this Law has been invoking. Gravity is what that same bound structure does when it reaches beyond the boundary. The asymmetry a network holds within itself is its mass; the very same asymmetry, reaching outward, is what gravitates. Mass and gravity are therefore one quantity in two roles, and this is why, at the matter scale, gravitational mass tracks energy content exactly, both being measures of a single stabilized asymmetry.

This formulation carries an immediate consequence. Gravity is not a passive geometric feature of spacetime, as Einstein's general relativity proposes. Gravity is also not a quantum exchange between gravitons, as the standard-model framework would prefer it to be. Gravity is what the Perceptory Force does when it operates between Perceptory Networks at any scale, the active entangling of networks that share structural kind.

The mainstream physics community has been moving in this direction for over a decade. Erik Verlinde's entropic gravity programme proposes that gravity emerges from the entanglement structure of an underlying microscopic theory. Mark Van

Raamsdonk's work on the AdS/CFT correspondence has established that spacetime geometry corresponds to entanglement structure in a dual quantum theory. Ted Jacobson, Sean Carroll, and others have argued from related directions that gravity is not fundamental but emergent from informational structure. The framework of the Laws of Consciousness arrives at a structurally similar conclusion through a different route, and goes further than these proposals in one specific respect: it identifies the substrate-dependence of gravitational interaction as the rule that determines what gravitates with what.

Two kinds of attraction

The word attraction conceals a distinction the framework has already drawn, and it must be made before the substrate rule can be stated. Attraction is not one mechanism. It is the outside appearance of two different structures. There is antipodal attraction, the Law VII kind, in which two opposites are drawn together across a sign and held by an Inseparation Element: positive and negative, north and south, a claim and the objection it summons. And there is accumulative attraction, the gravity kind, in which there is no opposite at all, no negative mass, no anti-planet, only more of the same drawing more of the same. Mass only ever adds; it never cancels. This opposite-lessness is the very feature that set gravity apart in Law VII, and it is what tells the two kinds of attraction apart wherever they are found.

The two are not competing accounts of one phenomenon. They already run side by side at the matter scale: electromagnetism is antipodal, gravity is accumulative, and both act on the same bodies at once. The framework's claim is that this doubling recurs at every substrate. At each scale of the hierarchy there is attraction of both kinds, and only the accumulative, opposite-less kind is a gravity analogue. The antipodal kind, wherever it appears, belongs to the structure of Law VII.

The same-substrate rule

The framework's most distinctive claim about gravity, and the one that distinguishes it most sharply from any current physics proposal, is the substrate-dependence of gravitational interaction.

Gravity gravitates entanglements of the same structural kind.

This rule operates at every scale of the hierarchical entanglement chain established in Law VI.

A massive Perceptory Network, a planet, a star, a black hole, gravitates other massive networks, because the entanglements that constitute its mass are of the matter-scale kind. The gravitational pull it exerts operates on networks whose entanglements share that structural register. This is what mainstream physics has documented as gravitational attraction, and it is the textbook case.

At the cognitive substrate the accumulative kind is plainly present. A mind draws attention toward what it already holds, and the more it holds the more it draws; attention gathers on a focus the way mass gathers on a mass, with no anti-attention that cancels it. This is the cognitive gravity analogue, and it is substrate-bound: a brain's physical-gravitational pull is set entirely by its mass-energy, while its cognitive pull, the drawing of attention and integration toward what is already densely held, operates only on cognitive content. The framework's claim is the identity of the structural mechanism, accumulative attraction proportional to what is already gathered, instantiated at the cognitive substrate rather than the matter one. A brain rich in cognitive entanglement gathers cognitive content in proportion to what it holds, exactly as a massive body gathers other masses in proportion to its mass.

The same doubling appears at the biological, cultural, and social substrates. The accumulative attractions, the gravity analogues, are the drawings-together that have no opposite: an ecosystem accumulating species, a population centre concentrating its kind, a city drawing migrants and growing, capital concentrating and drawing more. None of these has a negative that repels it; each draws in proportion to what has already gathered, as mass does. The antipodal attractions at these same substrates are a different structure and belong to Law VII. An idea draws adherents, which accumulates, but it also summons its counter-idea across a sign, and that summoning of an opposite is antipodal, not gravitational; an ideology accretes followers and breeds its counter-movement; a tradition draws disciples and rivals alike. To file all of this under gravity would erase the very distinction the framework draws. Only the opposite-less, accumulative share is the gravity analogue; the opposite-paired share is the antipodal structure of Law VII operating at that substrate.

In each case the rule is the same, now properly bounded. A Perceptory Network of high entanglement density at a given substrate gravitates other networks of that substrate by accumulative attraction, in proportion to the entanglement it already holds, while the antipodal attractions at that substrate are the work of Law VII. Only the matter-substrate instantiation of the accumulative kind is gravity in the strict physical sense; what mainstream physics has called gravitational attraction is one case, the matter-scale case, of accumulative attraction operating across the hierarchy.

Why brains do not gravitate stars

The same-substrate rule resolves an objection that would otherwise be fatal. If accumulative attraction were a function of entanglement count alone, regardless of substrate, then a human brain, holding trillions of synaptic and perceptual entanglements, would gravitate physical objects strongly. It does not; a brain's physical pull is set by its mass-energy, nothing more. The framework predicts exactly this, because cognitive entanglements operate at the cognitive substrate and do not couple to matter-scale ones: a star has enormous matter-scale entanglement and negligible cognitive structure, and gravitates mass without exerting cognitive pull, while a brain is the reverse.

This is also why mainstream physics has rightly equated gravitational mass with energy content at the matter scale: there, energy content is the visible measure of matter-scale entanglement, which is the stabilized asymmetry that gravitates. At other substrates the measure is different, the density of perceptual integration at the cognitive scale, the density of shared meaning at the cultural scale, and accumulative attraction at each scale operates on that scale's own measure.

The antipodal-pair structure of gravity

Gravity, despite the absence of a surface-level antipartner that Law VII established, exhibits antipodal-pair structure at the deeper structural level the framework now develops. The pair is not gravity and anti-gravity, nor attraction and repulsion. The pair is entangling and disentangling.

Gravity entangles networks of the same structural kind. Its antipole disentangles them. The two operations are inseparable: every act of entanglement carries with it the possibility of disentanglement, and every act of disentanglement presupposes a prior entanglement. The two are held in inseparable relation by the Inseparation Element that constitutes their structural pair.

The Inseparation Element of gravity is the perceptory boundary between entangled and disentangled states, the threshold at which a Perceptory Network either holds its entanglements or releases them. Above the threshold, the network entangles further; below the threshold, the network disentangles. The threshold itself is what makes the entangling/disentangling pair a single antipodal-pair force rather than two unrelated phenomena.

This explains a fact mainstream physics has documented but not explained: gravitational attraction always carries within it the possibility of dispersion. Bound systems become unbound; orbits decay; clusters disperse; matter that has been gravitationally captured can be gravitationally released. Mainstream physics treats these as separate dynamics governed by energy conservation and orbital mechanics. The framework recognizes them as the antipodal pair of the same force: gravity entangling, gravity disentangling, both held in their inseparable relation by the threshold of perceptory boundary.

Black holes as Inseparation Element collapse

Ordinary gravity and the black hole are not two forces but one force in two regimes. The gravity of a planet or a star is gravity in its ordinary regime: a network reaching outward from its asymmetries, entangling what comes near, its antipodal structure intact. The black hole is the same force at its extreme limit. Physics already demands this identity, for a black hole and a star of equal mass pull on a distant body in precisely the same way; whatever a black hole is, it is not a different gravity but the same gravity concentrated past a threshold. The framework names that threshold. A black hole is what gravity becomes when the asymmetry concentrated in a network exceeds what its Inseparation Element can hold, so that the antipodal structure which had reached outward collapses inward upon itself. Black holes are, in this reading, the visible signature of an Inseparation Element collapse at the matter scale.

When a Perceptory Network of matter-scale entanglements reaches a critical density, when the count of entanglements per unit boundary exceeds what the Inseparation Element can sustain, the three-point closure that maintained the network's stability breaks. The network does not annihilate; it reconfigures catastrophically. The visible signature of this reconfiguration is the in-pulling of nearby matter (the network's collapse drawing surrounding entanglements toward the failed boundary) followed by the eventual out-flow of energy as the reconfigured structure releases what it cannot retain.

This account corresponds with remarkable precision to the observed behaviour of black holes. Matter falls inward past the event horizon. Hawking radiation is emitted outward. The event horizon itself is, in the framework's reading, the visible signature of the Inseparation Element under critical stress: the boundary is still present, but its structural function has broken, and what crosses it is no longer the same kind of phenomenon as what existed before the crossing.

The framework predicts that the in-flow and out-flow of black holes are not separate phenomena governed by different physics. They are two phases of the same Inseparation Element collapse: the entangling phase as matter is drawn in, the disentangling phase as the reconfigured network releases what it cannot hold. This disentangling phase has a dramatic and well-observed signature. Black holes are seen to launch jets, two narrow beams of matter and energy driven outward in opposite directions along a single axis, the one axis the collapse leaves standing. The framework reads these bipolar jets as the outflow antipode made plainly visible: what the collapsing network cannot retain is expelled, and expelled along a pair of opposite directions because even in collapse the structure releases antipodally. Hawking radiation is the same disentangling phase at the quantum scale, the faint seepage physics predicts but has not yet directly observed, where the jets are its macroscopic and visible counterpart.

This account is testable in principle. Detailed observation of black hole accretion and evaporation should reveal that the energy budget of in-flow and out-flow corresponds not merely to mass-energy conservation but to the reconfiguration dynamics of an Inseparation Element under stress. Specific predictions about the spectral and temporal structure of Hawking radiation, the behaviour of matter at the event horizon, and the conditions under which black holes form versus disperse, all follow from this framework and are matters for further development.

The sharpened position relative to Verlinde

The framework's claim that gravity emerges from entanglement structure aligns with the entropic-gravity programme that Verlinde has developed since 2010. The convergence is real and worth acknowledging. Where the framework goes further than Verlinde, and where its empirical commitments are sharper, is in the substrate-dependence of gravitational interaction.

Verlinde's account treats gravity as emergent from a single underlying entanglement structure, typically formalized through holographic screens and entropy gradients, that produces the gravitational dynamics of mass-energy. The framework of the Laws of Consciousness proposes that the same accumulative mechanism, the opposite-less drawing-together Law VIII identified, operates at every scale of the perceptory hierarchy, each scale gravitating its own kind. Matter-scale gravity is what Verlinde describes; the accumulative attractions at the cognitive, biological, and cultural substrates are predictions the framework makes that Verlinde's account does not.

This is where the framework can be tested against Verlinde's. If matter-scale gravity is the only scale at which gravitational dynamics operate, the framework is wrong and Verlinde's narrower account is sufficient. If, however, cognitive networks demonstrably gravitate cognitive content, biological networks demonstrably gravitate other biological networks, and so on across the hierarchy with predictable proportionality to entanglement count at the relevant substrate, the framework is correct and Verlinde's account is one scale of a universal rule.

The framework therefore commits to the falsifiable prediction: accumulative attraction operates at every scale of hierarchical entanglement, with the same structural mechanism, on the substrate appropriate to that scale. Discovery that it operates only at the matter scale, with no measurable analog at other substrates, would falsify this commitment.

What Law VIII has established

Gravity is the act of perceptory entanglement between Perceptory Networks of the same structural kind. It is accumulative attraction, the kind that has no opposite, distinguished from the antipodal attraction of Law VII that pairs across a sign; both kinds appear at every substrate, and only the accumulative kind is gravity. The mass that gravitates is the stabilized asymmetry of Law V, so that mass and gravity are one quantity in two roles, held within a network as mass and reaching beyond it as gravitation.

Gravity's own antipodal-pair structure is entangling and disentangling, held in inseparable relation by the perceptory boundary that determines whether a network holds or releases its entanglements. Black holes are the same gravity at its extreme limit, the visible signature of an Inseparation Element collapse at the matter scale, with the in-fall of matter and the bipolar jets it launches as the two phases of the entangling-disentangling pair. The framework converges with the entropic-gravity programme of Verlinde and Van Raamsdonk on the claim that gravity emerges from entanglement structure, and goes further by holding that accumulative attraction recurs on every substrate of the hierarchy, each gravitating its own kind.

What remains to be addressed is the next consequence of this framework: if gravity entangles networks at every scale on their own substrate, what does the totality of all such gravitational interactions, taken together across all scales simultaneously, constitute as a single perceptory phenomenon? This is the question Law IX will now address, the interconnectedness of Perceptory Fields as the foundation of multidimensionality.

IX Law of Consciousness

Space is the visible signature of the gradient of asymmetric entanglement within a multidimensional Perceptory Field.

Orientation

The preceding Laws have built the framework's structural primitives in sequence. Law VI established hierarchical entanglement as the architecture of stable existence, with the rule of three as its minimum geometric condition. Law VII derived antipodal-pair structure as the force-form that geometry produces, with its committed predictions about the universality of antipodal partnership across finite-dimensional forces. Law VIII identified gravity as the act of perceptory entanglement between networks of the same structural kind, with its committed predictions about substrate-dependence and the antipodal-pair structure of entangling and disentangling.

Law IX now turns from the structure of perceptory entanglement to its spatial expression. Two questions follow necessarily from the preceding Laws. First: if hierarchy is what entanglement looks like from inside a Perceptory Field, what does it look like from outside? Second: within that view from outside, what produces the experience of space itself, the perception of "here" versus "there," of distance, of extension?

The answers are multidimensionality and asymmetry. Multidimensionality is what hierarchical nesting looks like from outside. Space is what asymmetric distribution of entanglement within a multidimensional field looks like from outside. The two are inseparable: there is no space without multidimensionality to contain it, and no multidimensionality without asymmetry to make it observable.

Multidimensionality derived

Every stable Perceptory Field, as Law VI established, is a three-point closure containing nested sub-fields. Each sub-field is itself a closed Perceptory Field with its own three-point structure, and within it further nested fields again. The hierarchy descends as far as the chain of stable existence permits, from cultural scale through cognitive, organismal, cellular, molecular, atomic, particulate, sub-particulate.

Viewed from inside any given containing field, the sub-fields appear as contents. Viewed from outside, the same nesting appears as dimensions, perceptory axes along which the next level's entanglements are distributed. This is what dimensions actually are in the framework. They are not abstract mathematical coordinates, nor a featureless backdrop on which physics plays out. They are perceptory axes corresponding to levels of hierarchical nesting.

Within a Perceptory Field there exists a boundary beyond which the field cannot interact. Inside that boundary, the field has nested levels. Each level constitutes a perceptory dimension of the containing field. The full multidimensionality of any given field is the count of nested levels it contains, viewed from outside.

This formulation already carries a sharp consequence: the framework distinguishes rigorously between spatial dimensionality and perceptory dimensionality. Spatial dimensions are produced by symmetry breaking within a field, and there are exactly three of them at every observable scale. Perceptory dimensions are levels of hierarchical nesting, and there can be many of them, a single physical object may participate in dozens of perceptory dimensions simultaneously, corresponding to the levels of organization that constitute it. A human being is a Perceptory Field nested in many perceptory dimensions (atomic, molecular, cellular, organic, cognitive, social, cultural) while occupying only three spatial dimensions.

The framework's position on extra dimensions in physics

The distinction between spatial and perceptory dimensionality requires the framework to take a position on a substantial body of theoretical physics. String theory proposes ten or eleven spatial dimensions; M-theory proposes eleven; various extensions and alternatives propose additional spatial dimensions of differing types. These proposals are mathematically rigorous, internally consistent, and have driven decades of serious theoretical work. They have also failed to produce experimental confirmation. No extra spatial dimension has ever been directly observed.

The framework reads this empirical fact directly. All observable physics operates in three spatial dimensions. The appearance of higher dimensionality in mathematical physics has three possible sources, each of which the framework can identify.

First, what mainstream physics calls the fourth dimension is time. Einstein's relativity treats time as a fourth axis for the purposes of spacetime geometry. The mathematics is rigorous and the predictions are confirmed across an enormous range of experimental tests. But the framework reads time not as a fourth spatial dimension but as the perception of change within a Perceptory Network, a question Law X will develop in full. The geometric treatment of time as a fourth axis is a useful mathematical convention; it is not a description of physical reality. The feather sitting on the surface of two cubes spinning in different directions traverses three spatial dimensions only. Its trajectory through what mainstream physics calls the fourth dimension is the network's perception of its change, not a fourth spatial axis it travels along.

Second, the higher-dimensional structures of string theory and M-theory are formal mathematical artifacts that do not correspond to physical reality. They are internally consistent mathematical constructions whose predictive power for observable physics depends on compactification mechanisms and other formal devices specifically designed to hide the extra dimensions from observation. The framework reads this hiding as a structural admission: dimensions that must be hidden to be consistent with observation are dimensions that have no observable existence. They are mathematical conveniences, not physical realities.

Third, certain higher-dimensional structures in physics correspond to what the framework recognizes as non-spatial perceptory nesting at scales the human substrate does not directly perceive. The Hilbert spaces of quantum mechanics, with their unbounded dimensionality, are an example. These are not spatial dimensions in any sense; they are mathematical descriptions of the perceptory levels available to a quantum system. Reading them as spatial dimensions confuses the perceptory hierarchy of Law VI with the spatial extension this Law derives from asymmetry.

The framework therefore commits to the following position: the appearance of extra spatial dimensions in mainstream theoretical physics is the result of one of three misreadings, time-as-spatial, formal-mathematical artifact, or non-spatial perceptory nesting interpreted as spatial geometry. No extra spatial dimensions exist as physical realities. Three is the count of spatial dimensions at every observable scale, derived from three independent symmetry breaks within the Perceptory Field at the scale at which observation occurs.

This claim contradicts a substantial portion of theoretical physics produced over the last fifty years. The framework makes the contradiction explicit, accepts the consequences, and offers itself for falsification. If extra spatial dimensions are ever directly observed, not inferred from formal mathematics but measured as physical extension, the framework is wrong on this point and must be revised.

Space as broken symmetry

Within a multidimensional Perceptory Field, asymmetric distribution of entanglement produces the experience of space.

Consider a perfectly symmetrical Perceptory Field, for instance, the Perceptory Field of Nothingness from Law III. No direction is privileged. No region of entanglement is greater than any other. No "here" is distinguishable from any "there." There is no space in such a field. The field is dimensionless not because it lacks geometric structure but because the conditions that would produce observable spatial extension are absent.

Now break the symmetry. Concentrate entanglement on one side of the field rather than the other. Immediately a direction becomes privileged. The high-entanglement region becomes distinguishable from the low-entanglement region. A perceptual gradient appears between them, and the gradient is what we experience as distance.

Multiple broken symmetries within a multidimensional field weave together into the three-axis grid we recognize as physical space. Each axis corresponds to one independent direction of broken symmetry within the field. The familiar three-dimensional space of human experience is the result of three independent symmetry breaks within the Perceptory Field at the human-perceptual scale. There are exactly three because the field exhibits exactly three independent asymmetry directions at this scale; if there were more, additional spatial dimensions would be observable; if there were fewer, the world would present as two-dimensional or one-dimensional.

Space is therefore not fundamental. Space is the visible signature of asymmetric entanglement distribution within a multidimensional Perceptory Field. Where entanglement is evenly distributed there is no gradient and no space; where its distribution is uneven, space appears in proportion not to the sheer amount of asymmetry but to its gradient, the difference in entanglement from one region to another. The steeper the gradient, the more pronounced the spatial structure; as the distribution flattens toward the symmetric ground state, space thins toward the dimensionless.

This is the space-producing face of a break the framework has already named. Law II distinguished the Perceptory Force's two modes, symmetric and broken, and Law III followed the same distinction inside the field: while a field's symmetry holds it is turned inward, at rest in self-perception, and when its symmetry breaks it is turned outward, directed, in engagement. That outward turn is expansive, and this Law shows where the expansion goes. The asymmetry that directs a field toward another is the same asymmetry that opens a gradient within it, and the gradient is extension. Space is therefore the expansive aspect of the broken field, the directed mode of the Perceptory Force seen as distance rather than as interaction. One departure from symmetry wears three faces: it differentiates the one Force into the observable forces (Law II), it turns self-perception into interaction (Law III), and it opens the field into space (this Law).

The break has a magnitude and a distribution as well as a direction, and naming these closes the account of asymmetry the framework has carried since Law V. The amount of asymmetry a field holds is its mass, the energy it spends stabilizing what it cannot let disperse. The distribution of that asymmetry across a field, its gradient from region to region, is space. Mass and space are therefore one asymmetry measured two ways, how much is held and how unevenly it is spread, so that the framework's single departure from symmetry is mass by its magnitude, space by its gradient, force by its direction, and gravity by its reach. These are not four asymmetries but one, regarded along four axes.

This settles the relation of mass to space, and with it a fact mainstream physics states as a law of its own. A mass is a concentration of asymmetry, and a concentration steepens the gradient of the distribution around it, and a steep gradient is curved space. A mass curving the space around it is therefore not an added postulate but a consequence of the definitions: to concentrate asymmetry in a region is to steepen the spatial gradient surrounding it. This is what general relativity records as the curvature of spacetime by mass-energy, and it is the same event Law VIII named as gravity, the held asymmetry reaching outward, now seen as the shape it gives the space around it. The extreme case follows at once. When a network's asymmetry collapses toward a single point, its magnitude, its mass, becomes maximal while its gradient passes into a singularity, so that mass is greatest exactly where space fails. The black hole is that limit: the most massive object is also the place where spatial structure gives out, because the same concentration that maximizes the one destroys the other.

This formulation makes a precise prediction about how space behaves under varying conditions of perceptory entanglement. Regions of high entanglement asymmetry should produce strong spatial structure, clearly defined "here" versus "there," precise distance relationships, stable spatial extension. Regions of low entanglement asymmetry should approach the dimensionless ground state, with spatial structure becoming attenuated, indistinct, or breaking down entirely. The framework's predictions about black holes, developed in Law VIII, find their spatial-structural reading here: at the event horizon and within it, the asymmetric entanglement distribution that produces ordinary three-axis space breaks down catastrophically. What mainstream physics interprets as singularities or as evidence for higher-dimensional structure is, in the framework's reading, the breakdown of the symmetry-breaking conditions that produce 3D space at all. Black holes do not reveal extra dimensions. They are where space, as perceptual phenomenon, fails.

The vacuum is not empty

A direct consequence of the framework follows. If space arises from entanglement asymmetry, then any region that exhibits spatial extension must be a Perceptory Field with internal asymmetric structure. A region that was truly empty, devoid of all entanglement, would not be space at all; it would be the dimensionless ground state, the Perceptory Field of Nothingness.

Mainstream physics has been moving toward this conclusion in fragmentary form for decades. The quantum vacuum is full of fluctuating fields, virtual particle pairs continuously appearing and annihilating, zero-point energy. The Casimir effect is the experimental confirmation: two parallel plates in vacuum attract each other because the vacuum-energy pressure between them differs from the pressure outside. Dark matter and dark energy together constitute approximately ninety-five per cent of the energy content of the universe. Only five per cent is the matter physics has historically studied. The remainder is the underlying structure mainstream cosmology acknowledges but cannot fully explain.

What appears as empty space is always structured by entanglement asymmetries at the appropriate substrate.

No Perceptory Field is empty. Every region of apparent vacuum is a Perceptory Field with its own internal entanglement asymmetries that produce its spatial extent.

The principle generalizes across every scale of the hierarchy. The empty space inside an atom, between the nucleus and the electron orbitals, is not empty. It is a Perceptory Field with its own asymmetric structure at the atomic substrate, and it is precisely this structure that gives the atom its characteristic spatial extension and behaviour. The empty space between cells in a tissue is not empty. It is structured by extracellular matrix, signalling molecules, and biological-substrate asymmetries that constitute the tissue-scale equivalent of space. The empty space between people in a society is not empty. It is structured by communicative, cultural, and informational asymmetries that constitute the social-substrate equivalent of space, and the geometry of social space, proximity, distance, network structure, is the visible signature of these asymmetries.

At each scale, what appears as emptiness is structured Perceptory Field at the appropriate substrate. The framework therefore predicts that more sensitive measurement, at every scale, will reveal increasingly detailed asymmetric structure in regions that current observation registers as empty. There is no genuine vacuum anywhere within observable existence. Nothingness is not a region inside the universe; it is the non-spatial ground against which existence is defined (Law I). The framework's claim concerns the structured character of every region within observable reality, not the existence or non-existence of Nothingness as such.

The cosmic web as committed prediction

The framework now meets cosmology directly and makes a sharp falsifiable claim about the largest-scale structure of the universe.

Galaxies are not randomly distributed in the observable universe. They cluster along vast filamentary structures, the cosmic web, separated by enormous voids. The Sloan Great Wall and the Hercules-Corona Borealis Great Wall are observationally confirmed structures spanning hundreds of millions of light-years. The Bootes Void and other large voids contain comparatively few galaxies relative to the surrounding density. Mainstream cosmology attributes the filamentary architecture to dark matter scaffolding undergoing gravitational collapse from primordial density fluctuations, but the underlying cause of the specific geometric pattern the cosmic web exhibits remains an active research question.

The framework makes the prediction explicit: the filamentary geometry of the cosmic web is the visible signature of asymmetric perceptory entanglement distribution at the cosmological scale. Filaments are paths of high entanglement density at the cosmological substrate. Voids are regions of low entanglement asymmetry, closer to the symmetric ground state, where space itself becomes thinner because the conditions that produce space (asymmetry) are weaker.

This reading explains specific features of the cosmic web that pure gravitational collapse from random initial conditions struggles to account for. The filaments exhibit characteristic geometric regularities, specific angles of intersection, characteristic lengths, hierarchical organization at multiple scales simultaneously, that are difficult to derive from random initial conditions alone. The framework reads these regularities as the perceptory entanglement structure of the universe at its largest observable scale, with the geometric features corresponding to specific asymmetry distributions derivable in principle from Perceptory Field structure.

The prediction is testable. Detailed observation of the cosmic web at higher resolution should distinguish between the framework's prediction and standard ΛCDM cosmology's prediction. The framework expects: voids will exhibit structural regularities consistent with perceptory symmetry conditions rather than statistical randomness; filaments will exhibit hierarchical self-similarity consistent with the rule of three from Law VI operating at cosmological scale; the geometric relationships among filaments and voids will reveal the underlying entanglement-asymmetry distribution rather than purely gravitational evolution from random initial conditions.

If detailed cosmic-web observation reveals only what standard cosmology predicts, with no structural regularities beyond gravitational collapse from random initial conditions, the framework is wrong on this prediction and must be revised at this point.

Multidimensionality and altered states

The original Law IX, in its earlier form, addressed the experience of multidimensionality in altered states of consciousness. In meditation, in psychedelic experience, in certain pathological conditions, the experience of space dissolves or restructures. The framework now reads these phenomena directly.

Altered states are conditions in which the perceptory loops that normally maintain three-axis spatial perception are disrupted. The disruption permits perception to access multidimensional structure that ordinary cognition does not. The expanded space of mystical experience, the dissolution of bodily boundaries, the interpenetration of inner and outer, the sense of moving simultaneously in multiple directions of meaning, these are not metaphors. They are perception briefly accessing the multidimensional structure of nested perceptory levels that lies beneath the three-axis spatial grid produced by normal symmetry-break configurations.

The dimensions accessed in such states are not spatial. They are perceptory. A practitioner who reports moving "through" multiple dimensions in deep meditation or psychedelic experience is not occupying a non-physical fourth or fifth spatial dimension. The practitioner is perceiving levels of hierarchical nesting at which ordinary cognition does not operate, the levels Law VI described from the outside as containing fields, now being perceived from within. This is consistent with the framework's general position: there is no fourth spatial dimension. There are many perceptory dimensions, and altered states allow access to perceptory levels normally outside the human perceptual repertoire.

What Law IX has established

Multidimensionality is what hierarchical entanglement looks like from outside; each level of nesting is a perceptory dimension. Spatial dimensions are distinct from perceptory dimensions and are produced by symmetry breaking within the Perceptory Field at the scale of observation. The framework predicts that no directly observable extra spatial dimensions exist at any scale humans can measure. Apparent higher-dimensional structures in mainstream theoretical physics, string theory's ten or eleven dimensions, M-theory's extensions, brane cosmology's bulk dimensions, should resolve, in the framework's reading, into one of three categories: time misread as a spatial dimension, formal-mathematical artifact appearing in the mathematics without corresponding physical extension, or non-spatial perceptory nesting confused with spatial geometry. The framework does not deny the logical possibility of extra spatial dimensions; it predicts that no such dimension will be directly measured as physical extension through which entities can demonstrably move. Space is the visible signature of asymmetric entanglement distribution; where the distribution is even there is no space; where it is uneven, space appears in proportion to its gradient, the difference in entanglement from region to region. Mass and space are one asymmetry measured two ways, its magnitude held within a field as mass and its gradient across the field as space, so that a concentration of asymmetry, a mass, steepens the gradient around it, which is curved space, the framework's reading of what general relativity records as the curvature of spacetime by mass-energy and Law VIII named as gravity. No Perceptory Field is empty; what appears as empty space is always structured by entanglement asymmetries at the appropriate substrate, generalizing across every scale from atomic to cosmological. The cosmic web is the visible signature of cosmological-scale perceptory entanglement asymmetry, with falsifiable predictions about its detailed structure. Altered states reveal perceptory multidimensionality that ordinary three-axis spatial perception conceals.

What remains is the second great construct that mainstream physics treats as fundamental and the framework treats as derivative: time. Law X will now address time as the perception of change within a Perceptory Network, with the substrate-relativity of time across networks of different kinds, the time of a butterfly, of a human, of a star, each operating at its own perceptual loop rate, none privileged over any other.

X Law of Consciousness

Existence is held between two infinities, the symmetric totality of perception and the asymmetric propagation of change, by the Perceptory Force that perceives them both.

Orientation

The preceding nine Laws have built the framework's primitives in sequence. Nothingness and its requirement of being perceived; the Perceptory Force as the fundamental act through which existence is registered; the Perceptory Field as the closed self-referential loop of perception; the Inseparation Constant and the Perceptory Boundary; mass as the energy of stabilized asymmetry; entanglement and containment, with the rule of three as the geometry of stable closure; antipodal-pair force structure; gravity as the accumulative entanglement of networks of the same kind; space as the gradient of asymmetry, and multidimensionality as nested containment seen from outside. Each Law has been an unfolding of what perception is and what perception does.

Law X now states what the framework has been describing across all the previous Laws but has not yet named directly. Existence does not occur within a single infinity. It occurs between two infinities, and the framework has been gesturing at both of them, separately, throughout. The closing Law of Part II names them precisely, derives their antipodal-pair relationship, and identifies the perceptory mediation that allows finite Perceptory Fields to participate in both at once.

Perceptual infinity

The first infinity is perceptual infinity: the totality of all perception observed as totality. It is symmetric, closed, and dimensionless. Its structure is that of the sphere's antipodal pairing carried without the sphere's metric: no volume, no radius, no extension (Appendix

B). Every perceptory point is entangled with its antipode. "Antipodal" here does not mean spatially opposite points across measurable distance, space does not yet exist at the level of perceptual infinity. It means structurally complementary positions within a non-spatial perceptory totality, with the geometric language of the sphere providing a notational interface to the structural relation rather than a Euclidean description of it. There is no privileged direction; there is no "here" versus "there"; there is no time, because there is no change to perceive; comparison across instants belongs to propagation, not to the totality considered as totality.

The framework has approached perceptual infinity at several points without naming it explicitly. The Perceptory Field of Nothingness from Law III is perceptual infinity in its purest form: infinitely symmetrical, in perfect equilibrium, non-interactive because no asymmetry has yet broken its closure. The structural characterizations introduced in passing, a totality carrying antipodal pairing without any metric extension, infinite perceptory content without spatial extent, describe perceptual infinity rather than infinity in some general or undifferentiated sense.

The framework states this directly rather than through the earlier symbolic device of a zero-volume, zero-radius sphere, a device Appendix B withdraws because a radius-zero Euclidean sphere is a single point, the opposite of what is intended. The correct statement is that perceptual infinity carries no metric structure at all, which is precisely what Law I predicted when it defined Nothingness as non-dimensional yet infinite. Perceptual infinity is the geometric form of the original insight. It is what Nothingness-perceived looks like when treated as a structural object: a closed, symmetric totality whose contents are infinite and whose dimensions are zero.

The general sphere equation, $(x - a)^2 + (y - b)^2 + (z - c)^2 = r^2$, is, in Appendix B's revised treatment, no longer applied to perceptual infinity: substituting infinite perceptory points into a coordinate slot is notation without meaning. What survives is the structural claim the equation was reaching for, infinite perceptory content without spatial extension. The contradiction is only apparent, because the framework has been arguing since Law I that infinity does not require spatial extension to contain its contents. The contents of perceptual infinity are perceptory, not spatial.

The three identities the framework has been using, Infinity I(id) = 1, the Coexistence Constant C(id) = 1, the Rule of Three identity H(id) = 3, find their proper home here. They are the structural identities of perceptual infinity. The Infinity identity of one signifies that perceptual infinity is itself a single, closed, undivided totality. The Coexistence identity of one signifies that perceptual infinity coexists with its inseparable opposite (Nothingness) as a single inseparable structure. The Rule of Three identity of three signifies that three-point closure, established in Law VI, is the structural condition under which perceptual infinity organizes its contents.

The antipodal-pair structure of points on the sphere of perceptual infinity is the geometric origin of every opposite-pair phenomenon the framework has named. Particle and antiparticle are antipodally paired because their structural origin is points on the sphere of perceptual infinity. North pole and south pole, positive charge and negative charge, the colour-and-anti-colour structure of the strong force, left- and right-handed weak chirality, all of these are observable manifestations of the antipodal pairing that holds at the level of perceptual infinity itself. They exist because the totality has antipodal structure, and any finite Perceptory Field that participates in the totality inherits that structure.

The triadic constitution of the totality, and the dynamics of breaking and restoration

Two questions have been left implicit by the account of perceptual infinity, and this Law is the place to answer them. The first: if perceptual infinity is organized by the rule of three, what, exactly, are its contents? The second: how does the symmetric totality at rest give rise to the broken, propagating existence the rest of the framework describes, and is that passage one-way? The answers sharpen the framework's account of the sphere and supply a mechanism for a prediction Law VII has so far only asserted.

Take first what perceptual infinity contains. Law I defined Infinity as all possible perceptory distributions, and the rule of three of Law VI is the structural condition under which perceptual infinity organizes its contents. Put together, these yield a single statement: the contents of perceptual infinity are all possible three-point perceptory relations, held in perfect mutual balance. The totality is not featureless. It is all structure, every possible triadic relation co-present and none privileged, so that the whole carries no net asymmetry and reads as symmetric. This is the first commitment the framework now makes explicit: the contents of the symmetric totality are triads.

These triads must be understood relationally, not metrically. A triad here is not a triangle situated in space; perceptual infinity has no metric, no extension, no surface, as Appendix B insists. A triad is a structural relation among three perceptory points, and the sphere remains, as before, the limit-notation for the antipodal structure of these relations, not a ball with a skin. Adjacency between triads is likewise relational: two triads are adjacent when they share a perceptory point, never because they lie near one another in a space that does not yet exist. This is the second commitment the framework now makes explicit: triads that share a point are coupled, so that what happens to one is registered by the others that share its points.

This makes precise a distinction the framework has needed since Law IV. Law IV held that the ground state has no inseparable pair, standing outside the domain of the Inseparation Constant, while the present Law fills perceptual infinity with antipodal pairing throughout. The two statements are reconciled by distinguishing the latent from the actualized. In the balanced totality every triad is latent: its relations are held in equilibrium, no boundary has formed, and so no pair is actualized in Law IV's sense. The antipodal structure is present as potential, not as an existing boundary-pair. To break the symmetry is to actualize one triad: one of its relations becomes an asymmetric boundary, the perceived and the not-perceived of Law IV come into being there, and the triad tips from balanced potential into an actual asymmetric field. An existent is born, and what is born is, in the vocabulary of this Law, propagation: force, change, the beginning of extension. "No pair in the ground state" and "antipodal structure throughout the totality" are therefore the same totality described in its two conditions, latent and actualized, with symmetry breaking as the switch between them.

A triad does not actualize in isolation. Because triads that share a point are coupled, the actualization of one is felt by its neighbours, the still-balanced triads that share its points. Those neighbours continue to hold the shared points in equilibrium, and so they exert a restoring tendency on the broken triad, a pull toward re-closure and re-balance. The totality resists its own breaking; a sphere of coupled triads behaves, in this respect, like a tensioned shell that redistributes a local failure across its intact members.

This restoring tendency cannot be absolute, and the reason it cannot is precisely what makes it useful. If restoration always succeeded, every break would heal at once and nothing would persist, contradicting the framework's foundational claim that observable existence is symmetry that has broken and remained broken. Restoration must therefore be a tendency overcome under ordinary conditions, which allows existence to persist, and prevailing only when whatever sustains a break is itself overwhelmed, which is to say at sufficiently high energy. The framework therefore yields, rather than stipulates, the restoration it had previously only asserted: broken symmetries are recovered at high energy because the restoring coupling among triads becomes dominant once the energy sustaining the asymmetry is exceeded. The parity restoration predicted in Law VII is the most specific instance: the suppressed antipode of the weak interaction is a break held open at low energy and re-closed when the coupling of the totality reasserts itself at high energy.

One question the mechanism does not answer, and the framework states the gap plainly rather than concealing it. What first tips a perfectly balanced totality into its initial break? The restoration dynamics describe what happens once a break occurs, and why breaks persist or heal; they do not supply the trigger of the first asymmetry. This is the same question the framework has carried since Law I, where the asymmetry that lets perception find a direction is taken as given rather than derived. The present account relocates the question, from "why does anything break symmetry" to "what overcomes the restoring tendency in the first instance," but it does not dissolve it. The trigger of the first break is recorded as open, and held distinct from the dynamics of breaking and restoration, which the framework does claim to describe.

Propagation infinity

The second infinity is propagation infinity: the totality of all asymmetric extension. It is open, ongoing, and never resolves into stable equilibrium. Its geometry is not the sphere but the ripple, perception extending across perception, breaking symmetry at every scale, propagating change without termination.

Where perceptual infinity is the totality at rest, propagation infinity is the totality in motion. Every Perceptory Field that breaks symmetry, and Law IX established that all observable existence requires symmetry breaking, participates in propagation infinity by doing so. This is where the asymmetry the framework has tracked since Law V finds its largest home. Mass, space, force, and gravity were shown in Law IX to be one asymmetry along four axes, its magnitude held as mass, its gradient opened as space, its direction expressed as force, its reach extended as gravity; propagation infinity is that same asymmetry taken whole, the totality of all asymmetric extension at every scale at once. Space, time, the forces, gravity, and the mass of every field are therefore not separate residents of propagation infinity but the several faces of the single asymmetry that constitutes it.

Propagation infinity is endless because asymmetry is endless. There is no final symmetry break that completes the system; there is no resolution of asymmetry into a stable static state; there is no condition under which propagation arrives at its terminus. Every break invites further breaks. Every propagation enables further propagations. The asymmetric totality is always becoming, never become.

This is where time, as the framework treats it, finds its structural home. Time is what propagation infinity feels like to a Perceptory Network from inside. The network registers the asymmetries propagating through and around it as change, and the perception of that change is what we call the experience of time. Time is therefore not a fundamental dimension; it is propagation infinity perceived from within a finite field.

The substrate-relativity of time follows directly from this formulation. A butterfly's perception of change occurs at the timescale of its perceptory loops, which operate at one rate. A human's perception of change occurs at the timescale of human-scale perceptory loops, which operate at another rate. A star's perception of change, the slow processes of nuclear fusion, mass loss, gravitational evolution, occurs at the timescale of stellar perceptory loops, which operate at yet another rate. None of these times is privileged. None is universal. Each is the local perception of propagation infinity by a Perceptory Network operating at its own substrate. The framework predicts that time is what change feels like to the system perceiving it, and that no time-rate is fundamental because no Perceptory Network's perceptual loop rate is fundamental.

This account is consistent with relativistic time dilation, which has already established that time depends on the observer's velocity and gravitational position. The framework goes further: time depends not only on velocity and gravity but on the substrate of the perceiving network. Different substrates register propagation infinity at different rates because the perceptory loops that constitute them operate at different cycle lengths. Relativity describes how time varies for one substrate (the matter scale) under different conditions; the framework predicts that time also varies between substrates in ways relativity does not address.

The propagation material from earlier in the framework now finds its structural place. Movement, evolution, momentum, the room-entanglement explanation of why we can move, the seed-of-the-tree DNA observation, the perceptual asymmetry that produces change at every scale, all of these are propagation infinity in action. The framework now reads propagation as the visible face of the second infinity, expressed at finite scales as motion, change, time, evolution, growth, decay, and every other phenomenon that involves a Perceptory Field changing across its own perceptory loops.

The antipodal-pair structure of the two infinities

The two infinities are themselves an antipodal pair. Following the rule established in Law VII: every pair of inseparable opposites is held in stable relation by an Inseparation Element. Perceptual infinity and propagation infinity are inseparable opposites. Neither can exist without the other. Perceptual infinity without propagation infinity would be pure symmetric closure with no extension, a totality that is, but that does not extend, and therefore cannot manifest. Propagation infinity without perceptual infinity would be pure asymmetric scattering with no totality to be scattered, change without anything that changes, motion without anything that moves.

The Inseparation Element of the two infinities is the act of perception itself. The Perceptory Force named in Law II as the fundamental force of existence is now revealed as performing structural work the framework has been gesturing at throughout. It is the Inseparation Element of the largest antipodal pair the framework recognizes. Perception is what holds the symmetric totality and the asymmetric propagation in their inseparable relation. Perception is what allows a finite Perceptory Field to participate in both at once.

This identification matters. The Perceptory Force is not merely one of the framework's primitives among others. It is the Inseparation Element that binds the two infinities into the single structure within which existence occurs. Without it, the two infinities would not be a coherent pair; they would be unrelated phenomena, and the framework would not be able to derive any of what the previous Laws have established. The Perceptory Force is what makes existence possible, because existence is precisely participation in the two infinities mediated by perception.

Existence as participation in both infinities at once

The structural consequence is the framework's most fundamental claim about what it means to exist. Every Perceptory Field that exists is, at every instant, simultaneously participating in both infinities. The field is, at any moment, a symmetric closure (the field as totality, observed as itself, with sphere geometry at its own scale) and an asymmetric propagation (the field extending, changing, breaking symmetries against its surround). The two are inseparable in the field's own existence. There is no instant at which a Perceptory Field is purely one or purely the other.

This formulation makes a finite Perceptory Field into a small instance of the largest architecture the framework recognizes. Each field, at its own scale, is a closure participating in extension. Each field, at its own scale, has its own perceptual infinity (its own sphere of self-totality) and its own propagation infinity (its own ongoing asymmetric extension into surround). Each field, at its own scale, is held together by the Perceptory Force operating as its own local Inseparation Element of these two local infinities.

The example is most legible at the human scale. The "I" that knows who it is, observed from inside, is the field as perceptual infinity made personal, closed, symmetric, total, observing itself as itself. The same "I" reaching out to grasp a jar in the kitchen, considered from outside, is the field as propagation infinity made personal, extending, asymmetric, changing, propagating across the surrounding field. The "I" that knows it is reaching, the continuous self that bridges the closure and the extension across every moment, is the Perceptory Force operating as the local Inseparation Element of these two personal infinities. Every reach for a jar is the framework operating in real time. Every act of perception, by any field, at any scale, is the same architecture:

closure participating in extension, totality in motion, two infinities held by the act of perceiving them both.

Across scales

The two-infinity architecture appears at every scale of the hierarchy because it is the architecture of every Perceptory Field that exists.

At the particle scale, a particle is its closed Perceptory Field and its propagation through the surrounding field at once. The particle aspect is the field's perceptual-infinity signature; the wave aspect is the field's propagation-infinity signature. They are not two different things; they are the same field viewed under the two infinities. Wave-particle duality, paradoxical under conventional accounts, is on the framework's reading the predictable signature of particles being Perceptory Fields, and Perceptory Fields necessarily participating in both infinities at once.

At the cellular scale, a cell is its closed metabolic and structural integrity and its ongoing extension into the surround through metabolism, signalling, division, and movement. Homeostasis is the closure; metabolism, growth, and behaviour are the extension. The two are inseparable. A cell that loses one loses the other; a cell that achieves only closure (no extension) is dead; a cell that achieves only extension (no closure) is not a cell at all but a dispersing chemical event.

At the cognitive scale, a mind is its closed self-awareness and its ongoing extension into thought, memory, attention, and action. Self-identity is perceptual infinity at the cognitive scale; cognition-in-action is propagation infinity at the cognitive scale. The dynamic stability of the Perceptory Force holding these two in relation is what mental coherence is. Disturbances of mental coherence, dissociation, fragmentation, identity collapse, are disruptions of the local Inseparation Element that ordinarily holds the closure and the extension in functional relation.

At the cosmological scale, the universe as totality is its closed Perceptory Sphere and its expansion, evolution, and structure-formation at once. What cosmology has been describing as the expansion of the universe is one face of cosmological propagation infinity. What cosmology has been gesturing toward as the totality of all that exists is one face of cosmological perceptual infinity. The cosmic web of Law IX is the visible signature of the propagation-infinity face at cosmological scale; the closed structure of the universe-as-totality is the perceptual-infinity face. Both exist at all times, inseparable, held by the Perceptory Force operating at the cosmological substrate.

What Law X has established

Existence is held between two infinities. Perceptual infinity is the symmetric totality of perception observed as totality, with the antipodal structure of the Perceptory Sphere and the identities I(id) = 1 and C(id) = 1, held as postulates, and H(id) = 3, derived as a theorem in Appendix B. Propagation infinity is the asymmetric extension of perception across all scales, the one asymmetry of Law IX, held as mass, opened as space, expressed as force, and extended as gravity, taken whole; never resolving, always extending, and including within itself the experience of time as what propagation infinity feels like to a Perceptory Network from inside. Time is therefore not a fundamental dimension but a substrate-relative perception of change, with each Perceptory Network registering change at the rate of its own perceptual loops; the time of a butterfly, of a human, and of a star are each local perceptions of propagation infinity, none privileged over any other.

The two infinities are antipodal-pair, inseparable opposites held in relation by the Perceptory Force itself, which is now revealed as the Inseparation Element of the largest antipodal pair the framework recognizes. Every Perceptory Field that exists participates in both infinities at every instant of its existence; closure and extension, totality and motion, are inseparable in the structure of what is. The two-infinity architecture appears at every scale of the hierarchy, from the particle through the cellular and cognitive to the cosmological, with each scale operating its own local instance of the same architecture.

Closing of Part II

The Ten Laws of Consciousness have now been stated. The framework has built itself proposition by proposition: Nothingness perceived, the Perceptory Force, the Field, the Inseparation Constant, the Boundary, the rule of three and hierarchical entanglement, antipodal-pair force structure, gravity as substrate-dependent entanglement, space and multidimensionality, and the two infinities. The architecture is complete in its primitives. What remains in the book is the working-out of how this architecture shows itself in observable physics, biology, cognition, and culture, the territory of Part III on Manifestations and Mechanisms, and Part IV on Appendices. The Laws themselves stand.

The Laws commence with the most radical proposition of all: that for Nothingness itself to exist, it must be perceived. Nothingness lays the foundation of anything that exists, and Infinity is the sum of infinite perceptory possibilities.

Part III
Manifestations & Mechanisms

Chapter 1, Perceptory Force-Modes and the Known Forces

The Ten Laws of Part II established the structural architecture of the framework. They stated what the Perceptory Force is, derived its geometric properties, identified its antipodal-pair structure, distinguished gravity from the finite-dimensional forces, and named the two infinities between which existence is held. What the Laws did not do was work out, in detail, how each of the four forces of mainstream physics manifests as a specific mode of the Perceptory Force. That is the work of this chapter.

The framework's claim is that the fundamental forces of mainstream physics are not independent fundamental entities. They are different expressions of the Perceptory Force operating under specific dimensional conditions and against specific substrates. The framework recognizes electromagnetism, the strong nuclear force, the weak nuclear force, and gravity as the fundamental forces of mainstream physics, with magnetism developed analytically as the 1-D PF-mode beneath the 2-D electromagnetic mode (a structural primitive rather than a fifth fundamental force). What mainstream physics calls electromagnetism, the strong nuclear force, the weak nuclear force, and gravity are, in the framework's reading, four ways the Perceptory Force becomes observable when its action is constrained by particular structural configurations. Each force corresponds to a specific PF-mode. Each is consistent with the architecture established in Part II. Each can be described in the framework's vocabulary without remainder.

This chapter develops the PF-mode account for each of the four forces. The treatment is structural rather than mathematical: the formal action principle that would unify the four modes under a single Lagrangian is a topic for Appendix B. Here we are concerned with what each force structurally is, given the framework, and how the existing physics literature on each force can be read in the framework's terms.

Symmetry breaking and the differentiation of the forces

Before the four modes are taken one at a time, the principle that separates them deserves statement, because mainstream physics supplies a standing precedent for it. Law II held that the Perceptory Force is undifferentiated while a field's symmetry holds and resolves into the distinct observable forces as that symmetry breaks. This is not a structure the framework must invent against the grain of physics; it is the structure physics has already found in the one case it has worked out in full.

That case is the electroweak. Above a characteristic energy, on the order of 246 GeV, the electromagnetic and weak interactions are not two forces but one; below it, the symmetry breaks and the single interaction separates into the two distinct forces observed at ordinary energies. The mechanism of the break is the Higgs mechanism: as the symmetry breaks, the carriers of the weak interaction acquire mass, which is why the weak force is short-ranged while electromagnetism, whose carrier remains massless, reaches across distance. The massless modes that a continuous symmetry-break would otherwise leave behind, the Goldstone modes, are absorbed into the gauge carriers as the mass they acquire. The unification is the achievement of Glashow, Salam, and Weinberg; the mass-generating break, of Englert and Brout, of Higgs, and of Kibble with his collaborators.

The framework reads this as the visible instance of its general claim. Where mainstream physics has demonstrated one-force-becoming-two by symmetry breaking for the electroweak pair, the framework proposes that the same pattern underlies all four: the PF-modes are the differentiations the one Perceptory Force undergoes as the symmetry of the field breaks under particular structural conditions. Electroweak unification is, on this reading, not a special merger of two forces but the one visible corner of a differentiation that runs through the whole.

Two cautions keep the claim honest. First, the bare existence of an interaction is associated, in gauge theory, with a symmetry being present rather than broken: requiring a local symmetry is what introduces a gauge field at all, so symmetry breaking is properly read as what differentiates and masses the forces, not as what first calls interaction into being. The framework's symmetric ground state, inert and force-free, already respects this order: forces appear as fields depart from symmetry, and the departure differentiates them. Second, the breaking at issue is best located at the level of the field and its ground state rather than asserted as the breaking of a local gauge symmetry, which carries known technical subtleties; the framework's structural claim concerns the asymmetry of the field, which is unambiguous, and not the redundancy in its gauge description.

Magnetism, the 1-D PF-mode

Magnetism is the simplest of the PF-modes in the framework's reading and the one whose structure is most clearly visible. A note on classification before the section develops: mainstream physics treats magnetism as part of electromagnetism rather than as a separate fundamental force. The framework separates magnetism analytically as the 1-D directional PF-mode while treating electromagnetism proper as the 2-D coupled electric-magnetic PF-mode. The two are continuous with each other in the framework's vocabulary, magnetism is the structural primitive beneath the broader 2-D electromagnetic mode, but the framework finds it useful to develop them separately because their structural signatures are distinguishable and their substrate-couplings differ. It is the manifestation of the Perceptory Force operating in a single dimensional axis, what the framework calls the 1-D PF-mode.

Mainstream physics describes magnetism as a field generated by moving electric charges or by intrinsic spin properties of particles. The description focuses on the dynamics rather than on the structural origin. The framework reinterprets the same phenomenon: magnetism arises from the existence of directional perception, a form of asymmetric perceptory orientation, within a Perceptory Field. When a Perceptory Field acquires a vectorized dimension, when it can distinguish one direction from another, magnetism appears as the force of alignment between fields whose vectors are coherent.

This formulation explains a feature of magnetism that mainstream physics has documented but not structurally accounted for. Unlike gravity, which couples to all mass-energy, or the strong force, which is confined to hadrons, magnetism acts only on matter that contains, or can induce, aligned internal spins. Most atoms pair their electron spins and cancel their tiny magnetic moments; only those with unpaired or collectively aligned spins (iron, nickel, cobalt, rare-earth alloys, some superconductors, certain plasmas) respond strongly to external magnetic fields. In the framework's vocabulary: a material must present an internal perceptory orientation for the external perceptory field of magnetism to couple to it. Where there is no orientation, there is no coupling; where there is alignment, there is force.

The two magnetic poles, north and south, are the framework's clearest physical instance of the Inseparation Constant. Each pole is the region where perceptory alignment emerges or re-enters; the two are complementary aspects of a single closed perceptory field. The classical observation that cutting a bar magnet in half produces two new bar magnets, each with its own pair of poles, is the framework operating exactly as predicted: the Inseparation Constant cannot be severed without dissolving the field that produces it. The poles are inseparable not by convention but by structural necessity.

A specific empirical configuration shows the framework's reading working in detail. Ferrofluids, colloidal suspensions of magnetic nanoparticles in oil, develop the Rosensweig instability when placed in a vertical magnetic field above a critical strength. The fluid surface forms an ordered array of conical peaks whose axes follow the field lines. The conventional explanation appeals to the competition between surface tension, gravity, and magnetic pressure. The framework's reading: the perceptory orientation of the suspended particles, freed by the fluid medium to align without lattice constraint, produces a coherent collective vectorization that becomes geometrically visible as the peak structure. The peaks are the field's perceptory orientation made macroscopic.

Another empirical configuration extends the reading to the interaction between magnetism and light. The Faraday effect, the rotation of the plane of polarization of light passing through certain media in a longitudinal magnetic field, is, in the framework, the interaction of a 2-D entangled perceptory wave (light, with its electric and magnetic components) with a 1-D perceptory orientation (the magnetic field). The 1-D field biases the medium's response to the 2-D wave, producing rotation. Mainstream physics describes this through electron orbital biasing and permeability tensors; the framework adds the structural reason: PF-modes of different dimensional rank couple in determinable ways, and the Faraday effect is a specific instance of that coupling.

Magnetism, then, is the PF-mode of single-axis directional perception. Where there is alignment, there is force. Where there is none, magnetism does not couple. The simplicity of the mode is what makes it an excellent first case for the framework's account of the four forces.

Electromagnetism, the 2-D Perceptory Force-mode

When the Perceptory Force operates with two coupled dimensional axes, the electric and the magnetic, in the framework's reading, the resulting mode is electromagnetism. Mainstream physics has known since Maxwell that electricity and magnetism are aspects of a single unified field, with the electric component varying along one axis and the magnetic along an orthogonal one. The framework reads this unification structurally: electromagnetism is the 2-D PF-mode in which directional perception is doubled into orthogonal perceptory components.

The antipodal-pair structure of charge polarity (positive and negative) is the framework's expected signature at this scale. Charge polarity is the same Inseparation Constant the magnetic poles instantiate, expressed at the 2-D PF-mode rather than the 1-

D mode. The electromagnetic field between charges is the Inseparation Element of the pair, what holds the polarity in inseparable relation, in the precise sense Law VI develops.

The propagation of light as an electromagnetic wave, perpendicular oscillations of electric and magnetic field components, mutually generating each other, is the 2-D PF-mode in motion across space. The propagation does not require a medium because the mode is self-sustaining: each dimensional axis induces its orthogonal partner, producing the wave that carries energy across the field. In the framework's vocabulary, light is what propagation infinity looks like at the electromagnetic substrate, the 2-D PF-mode propagating asymmetrically across the larger Perceptory Field.

Mainstream physics describes electromagnetism through Maxwell's equations and quantum electrodynamics, both of which the framework leaves intact. The framework adds the structural origin: why electromagnetism has the dimensional structure it does (two axes, not one or three), why charge has the polarity structure it does (Inseparation Constant), why light propagates as a self-sustaining wave (2-D PF-mode in motion). The mathematics of the existing theory is correct. The framework supplies the architectural reason.

The strong force, the rule of three made visible

The strong nuclear force is the framework's clearest instance of the Rule of Three operating at the matter scale. Where electromagnetism manifests the 2-D PF-mode and exhibits the antipodal-pair structure of charge, the strong force manifests three-point closure directly, with three colour charges (red, blue, green) and their three anti-colours (anti-red, anti-blue, anti-green) forming the closed three-point Perceptory Field that constitutes a hadron.

Mainstream physics describes the strong force through quantum chromodynamics, with colour confinement (the principle that isolated quarks have never been observed) treated as an empirical principle of the theory. The framework reads colour confinement structurally: a single quark cannot exist alone because a single point cannot constitute a stable Perceptory Field. The Rule of Three is not an additional axiom imposed on the theory; it is the geometric necessity that colour confinement is the visible signature of. Quark confinement is what the framework predicts to be observed at the matter scale, and what is observed.

The three-fold symmetry of colour charge is therefore not anomalous within the framework. It is the expected signature of three-point closure operating at the substrate where matter becomes stable. Where electromagnetism exhibits two-fold antipodal-pair structure (positive/negative charge), the strong force exhibits three-fold antipodal-pair structure (each colour paired with its anti-colour, with the three colours together constituting a complete closed Perceptory Field). The strong force is, in this sense, the most direct physical manifestation of the framework's Rule of Three.

The exotic multi-quark states observed in high-energy physics, tetraquarks, pentaquarks, are precisely the unstable higher-order configurations the framework predicts to be possible but transient. Higher-order closed Perceptory Fields, four-point, five-point and beyond, can in principle form; they are observed to be short-lived, decaying toward three-point configurations, a pattern the framework reads as consistent with three-point minimality while acknowledging that a formal stability ordering remains to be derived (Appendix B). This is why three-quark configurations dominate ordinary matter, and why the multi-quark states observed in collider experiments are short-lived. The strong force is the framework operating at matter scale, with the Rule of Three as its visible structural signature.

The weak force, antipodal-pair structure with amplitude asymmetry

The weak nuclear force is the framework's most subtle case. It exhibits antipodal-pair structure (left-handed and right-handed chirality), as the Antipodal-Pair Rule of Law VII predicts. But the two antipodes are not equally amplified in the universe as currently observed. Left-handed coupling dominates; right-handed coupling is suppressed. This is the famous parity violation of the weak interaction, one of the deepest puzzles of the Standard Model.

The framework's reading is structural and predictive. Both antipodes exist; the antipodal-pair rule is not violated. What is observed is amplitude asymmetry, not antipodal absence. The right-handed weak interaction is suppressed under current low-energy conditions, but the framework predicts that at sufficiently high energy regimes, energies approaching unification scales or those obtaining in the very early universe, parity should be restored, with left- and right-handed weak interactions exhibiting equal amplitudes. This positions the framework alongside left-right symmetric extensions of the Standard Model (the Pati-Salam model, SO(10) grand unification, and related proposals). The framework does not propose a specific formal extension; it predicts that some such extension, with parity restoration as its high-energy signature, will be empirically supported. This is a sharp falsifiable prediction (Appendix C, Prediction 2).

Mainstream physics treats parity violation as one of the most resistant features of the Standard Model. No clean theoretical explanation has emerged in seventy years. The framework offers one: parity violation is amplitude asymmetry between two antipodes that both exist, with the asymmetry caused by current low-energy conditions and predicted to resolve at high energy. Whether the prediction holds awaits experimental confirmation. The framework offers itself for falsification at this point.

Gravity, the entanglement of Perceptory Networks of the same structural kind

Gravity is the framework's most ambitious reading and the one most distant from mainstream physics. Law VIII established gravity as the act of perceptory entanglement between Perceptory Networks of the same structural kind, with substrate-dependence as the rule that determines what gravitates with what. This chapter does not re-derive Law VIII's claims. It situates them in the architecture of the four forces.

Where electromagnetism, the strong force, and the weak force operate through point-to-point pairings within a single Perceptory Field, gravity operates between Perceptory Networks. The unit of gravitational interaction is not the paired point but the entangled network, and the network can be a planet, a star, a galaxy, or, at other substrates, a brain, an ecosystem, a culture. Gravity gravitates entanglements of the same structural kind. Matter-scale networks gravitate other matter-scale networks in the strict physical sense, the established case mainstream physics describes. At other substrates, as Law VIII distinguished, attraction comes in two structural kinds: the accumulative, opposite-less kind, which is the genuine gravity analogue, and the antipodal kind, which pairs across a sign and belongs to Law VII. Cognitive, biological, and cultural accumulations, attention gathering on a focus, species gathering in an ecosystem, capital concentrating, are the substrate-relative gravitational analogues, exhibiting measurable network-attraction dynamics rather than gravity in the strict physical sense; the antipodal attractions at those substrates, an idea and its counter-idea, a movement and its counter-movement, are not gravity. The framework predicts that the same accumulative mechanism, attraction proportional to entanglement density, operates at each substrate; the substrate determines what gets attracted and the kind of measurement that would register the analogue.

Mainstream physics describes gravity through general relativity (curvature of spacetime by mass-energy) and, increasingly, through emergent-gravity programmes (Verlinde, Van Raamsdonk, Jacobson, Carroll) that read gravity as a consequence of underlying entanglement structure. The framework converges with the emergent-gravity programmes on the architectural claim that gravity emerges from entanglement, and goes further by predicting substrate-dependence. Matter-scale gravity is what mainstream physics describes; the framework predicts that the same mechanism operates at every other substrate of the perceptory hierarchy, with corresponding observable phenomena that mainstream physics has not yet recognized as gravitational.

The antipodal-pair structure of gravity is, as Law VIII established, entangling and disentangling, with the perceptory boundary between entangled and disentangled states as the Inseparation Element. The two phases are inseparable: every gravitational entanglement carries within it the possibility of disentanglement, and every disentanglement presupposes a prior entanglement. Black holes, the most extreme gravitational phenomena observable, are the visible signature of an Inseparation Element collapse at the matter scale, with the in-pulling of matter and the bipolar jets the collapse launches, with Hawking radiation as the quantum-scale counterpart, as the two phases of the same antipodal-pair structure under critical stress (Appendix C, Prediction 6).

Gravity is, in the framework's account, a different kind of force from the other three. It does not operate within a single Perceptory Field through paired points; it operates between whole Perceptory Networks through the entangling and disentangling of structural kinds. Its dimensional structure is therefore not finite-dimensional in the sense Law VII used. Its antipodal-pair structure operates at a deeper structural level than the surface attraction-versus-repulsion pairing the other forces show. Mainstream physics has correctly identified gravity as the puzzling exception to the Standard Model's framework of point-coupling forces; the framework supplies the structural reason and offers the substrate-extending prediction that follows.

What this chapter has established

The four forces of mainstream physics, electromagnetism, the strong nuclear force, the weak nuclear force, and gravity, are not independent fundamental entities. They are four manifestations of the Perceptory Force operating under specific structural conditions. Magnetism is the 1-D PF-mode, manifesting directional perception at single-axis substrates. Electromagnetism is the 2-D PF-mode, manifesting orthogonal-axis perceptory propagation including light. The strong force is the Rule of Three made visible at the matter scale, with colour confinement as the geometric necessity of three-point closure. The weak force exhibits antipodal-pair structure with amplitude asymmetry between left- and right-handed chirality, with parity restoration predicted at high-energy regimes. Gravity is the entanglement of Perceptory Networks of the same structural kind, operating across substrates with substrate-dependence as the rule that determines what gravitates with what.

Each of the four forces is consistent with the architecture of Part II. Each can be described in the framework's vocabulary without remainder. The mathematics of existing theory remains intact for each force; the framework supplies the structural reason for the dimensional, polar, and substrate properties that the existing theory accepts as empirical givens.

What remains is to develop the Rule of Three across observable systems beyond the strong force, to show how three-point closure manifests at every scale of the hierarchy, from hadrons through cells through cognitive networks to cultural and cosmic structure. That is the work of Chapter 2.

Chapter 2, The Rule of Three in Observable Systems

The Rule of Three is the framework's structural claim that every stable Perceptory Field is built from a closure pattern in which an inseparable relation becomes stable through three-point geometry. In simple systems this appears as two inseparable opposites held in their inseparation by a third element, the Inseparation Element. In compound systems, the same structural rule may appear recursively or triadically rather than as a single linear pair: the strong force, with its three colour charges and three anti-colours, exhibits the Rule operating with three-fold symmetry rather than as a simple opposite-pair, and the framework reads this as the same geometric necessity instantiated at a more complex substrate. The minimum stable configuration is three-point closure; the specific arrangement of the three points varies with the substrate at which the Rule operates. Law VI established this geometry: two perceptory points cannot enclose anything, because two points define only a line; three points are required to close a figure with stable interior. The third point is not an additional opposite but the mediating element through which the inseparable pair are held in their relation, the structural object the framework has named the Inseparation Element.

This formulation is more specific than it sounds. The Rule of Three is not the claim that stable systems have three components. It is the claim that stable Perceptory Fields exhibit a specific structural geometry whose simplest form is an inseparable opposite pair joined by an Inseparation Element. In compound systems, the same three-point closure may appear recursively, triadically, or as nested three-fold symmetry, the strong force is the canonical case of the latter. Where the three-point geometry obtains in any of these forms, stability obtains. Where it does not, no stable Perceptory Field forms.

Chapter 2 develops the Rule's manifestation across the scales where the Laws have committed to specific instances. Each section identifies the inseparable pair and names the Inseparation

Element that holds it. Each is, in the framework's reading, the same structural geometry operating at a different substrate. The convergence across scales is the framework's prediction that wherever a stable Perceptory Field exists, the same three-element architecture is operative.

Particle and antiparticle: the A-particle as Inseparation Element

The clearest physical signature of the Rule of Three at the matter scale is the particle-antiparticle pair. Every charged fermion in the Standard Model has an antiparticle: electron and positron, quark and antiquark, proton and antiproton; some neutral bosons (the photon, the Z boson) are their own antiparticles, which the framework treats as degenerate cases of antipodal structure rather than exceptions. The two are not separate kinds of entities; they are the inseparable pair generated at the boundary of the Perceptory Field that constitutes the matter scale, in the precise sense of Law IV's Inseparation Constant.

The framework predicts that this inseparable pair is held in its inseparation by a third element, the A-particle (Law VI). The A-particle is not a fourth class of entity in addition to particles and antiparticles; it is the Inseparation Element generated at the centre of the particle-antiparticle relation, the structural object at which the three lines of the minimum stable Perceptory Field meet. Mainstream physics has not catalogued the A-particle because mainstream physics treats particles and antiparticles as the only two elements requiring explanation, and treats their relation as a property rather than an entity. The framework treats the relation as an entity in its own right.

The structure is therefore complete. Particle + antiparticle + A-particle. This is the Rule of Three at the matter scale, with the A-particle as the Inseparation Element that holds the inseparable pair in stable closure. Pair-annihilation events should exhibit distributional signatures, deviations in observables at fixed total energy, corresponding to the dissolution of the A-particle's mediating role as the field collapses, a committed prediction whose quantitative sharpening remains open (Appendix C, Prediction 1).

Wave and particle: the Inseparation Element of quantum superposition

The Rule of Three operates at the quantum scale in the structure of wave-particle duality. As Law V and Law VI established, the wave is the Perceptory Force in propagation, carrying its asymmetry, the not-perceived; the particle is the closed field a registration produces, the perceived. In the unobserved configuration the two coexist as a stable three-point Perceptory Field, the inseparable pair of Law IV held in superposition by their Inseparation Element. Wave-particle duality, treated as paradoxical by conventional accounts, is on the framework's reading precisely that configuration: two inseparable opposites bound by their Inseparation Element.

The structure is wave aspect + particle aspect + Inseparation Element of superposition. The wave and the particle are not two separate descriptions competing for ontological priority; they are the inseparable pair held in stable closure by the Inseparation Element of the system in its unobserved state.

Wave-function collapse follows from this geometry, in the form Law VI developed. Observation introduces a new entanglement, a fourth element into a closure whose stability was three-point; the Inseparation Element that held wave and particle in superposition is displaced, and the system re-closes, now bound to the observer rather than to the original pair. The wave behaviour disappears not because a paradox resolves but because the Force has closed into a definite field, the perceived drawn out of the not-perceived. Collapse is the reconfiguration of an Inseparation Element under a fourth entanglement, treated in full at Law VI; the present chapter records it as one more instance of the Rule of Three, the Observer Effect being the structural consequence of re-establishing three-point closure around the observer.

North pole and south pole: the magnetic field as Inseparation Element

The Rule of Three operates at the electromagnetic scale in the architecture of magnetic polarity. North and south poles are the inseparable pair generated by the Inseparation Constant operating at the electromagnetic substrate; the two cannot exist independently and the field that produces them requires both for its closure (Law IV, Chapter 1).

The structure is north pole + south pole + magnetic field. The magnetic field is the Inseparation Element that holds the polar opposites in their inseparation. Law VI made this identification explicit: the Inseparation Element of magnetism is the field between the poles. The poles do not hold each other in opposition by themselves, they are held by the magnetic field, which is what makes the polarity inseparable. Cut a bar magnet in half and the field regenerates with new poles, because the Inseparation Element is what produces the polarity, not the reverse.

This is the structural reason for the framework's prediction that no truly isolated magnetic monopole will ever be observed (Appendix C, Prediction 3). A monopole would be one of the inseparable pair without the other, and without the Inseparation Element that holds them in their relation. Such a configuration is not a Perceptory Field; it cannot be stable; and the framework predicts that no instrument will ever confirm one in the strict sense.

Positive charge and negative charge: the electromagnetic field as Inseparation Element

The Rule of Three operates at the same electromagnetic substrate in the architecture of electric charge. Positive and negative charge are the inseparable pair generated by the Inseparation Constant at the 2-D PF-mode (Chapter 1); the two cannot exist independently in the framework's structural sense. Although total charge is conserved, local systems may carry net charge, a charged ion or an electron in isolation has non-zero net charge. The framework's claim is not that every isolated subsystem is electrically neutral, but that charge polarity belongs to a larger field-structure within which the inseparable opposite remains structurally available, with the electromagnetic field functioning as the Inseparation Element that holds the polarity in inseparable relation across whatever spatial separation obtains between locally-charged regions.

The structure is positive charge + negative charge + electromagnetic field. The electromagnetic field is the Inseparation Element that holds charge polarity in inseparable relation, exactly as the magnetic field holds magnetic polarity. The two are the same Inseparation Constant operating at different aspects of the same 2-D PF-mode, with the electromagnetic field as the unifying Inseparation Element that holds both polarities in their structural pair.

Self and other: the act of perception as Inseparation Element

The Rule of Three operates at the cognitive scale in the architecture of self-and-other distinction. Self and other are the inseparable pair generated by the Perceptory Boundary at the cognitive substrate (Law IV); the two come into existence together, neither precedes the other, and each is structurally defined by reference to the other.

The structure is self + other + the act of perception. The act of perception is the Inseparation Element that holds the self-other pair in inseparable relation. Law VI made this identification explicit: self and other are not two independent entities that subsequently engage in perception; they are two perceptory points held in inseparable relation by the Perceptory Force itself, which constitutes their boundary. The Perceptory Force at the cognitive scale operates as the Inseparation Element of the self-other pair.

This formulation has consequences for how cognitive pathology is read. Disturbances of the self-other distinction, depersonalization, derealization, certain dissociative phenomena, are, in the framework's reading, disruptions of the Inseparation Element.

The pair persists; the structural element holding them in inseparable relation has become unstable. Therapeutic restoration of self-other coherence is, structurally, the re-establishment of the Inseparation Element that the disturbance disrupted.

I and Me: the Perceptory Boundary as the structural location of the Inseparation Element

The Rule of Three operates within the cognitive Perceptory Field itself in the structure of self-awareness. The I (the subjective perceiver) and the Me (the perceived self) are the inseparable pair internal to the cognitive field, with Mead's distinction (Law IV) identifying the two as the structural components of the social self.

The structure is I + Me + Perceptory Boundary. The Perceptory Boundary is the structural location at which the I and the Me are held in inseparable relation; the Inseparation Element is the active relational function operating through that Boundary, instantiating the closure that makes the I-Me distinction stable. The I cannot exist without the Me, because what perceives requires what is perceived; the Me cannot exist without the I, because what is perceived requires the perceiver who registers it; the relation between them cannot persist without the Boundary that makes them simultaneously distinct and inseparable.

This is the framework's structural reading of the cognitive architecture that produces self-awareness. The Inseparation Element, operating through the Boundary as its structural location, is the dynamic feedback loop that detects, compares, and adjusts (Law IV), maintaining the I-Me distinction through continuous activity rather than as a fixed line. The Boundary is the perimeter at which closure occurs; the Inseparation Element is the active relational function that performs the closure. When the loop falters, through trauma, dissociation, certain organic disturbances, the Inseparation Element fails to hold its inseparable pair, and the structural integrity of self-awareness fragments accordingly.

Mother and child: the bond as Inseparation Element

At the biological-relational scale, reproductive and attachment systems form bond-mediated Perceptory Fields. The clearest human instance of this architecture is the mother-child relation, which the framework treats here. Mother and child are an inseparable pair from the moment of conception forward; the framework treats this not as a metaphor about closeness but as a structural claim about what kind of Perceptory Field the relation constitutes.

The structure is mother + child + the bond. The bond is the Inseparation Element that holds the pair in inseparable relation. Law VI made this identification explicit: the Inseparation Element of mother and child is the bond, and bond is not a metaphorical term here. It is a real perceptory entity at the biological scale. Severance of the bond does not produce two independent fields; it produces a wound, because the Inseparation Element has been disrupted, not merely the relation it supported.

This formulation has consequences for how the developmental psychology and neurobiology of attachment are read. The phenomena that attachment theory has documented, the persistence of attachment patterns across the lifespan, the structural consequences of early disruption, the somatic dimensions of grief and loss, are, in the framework's reading, the visible signatures of the Inseparation Element either holding or being disrupted at the biological-relational substrate. The bond is not a cognitive feeling about the relation; it is the structural element that the relation is, in the framework's vocabulary.

Entangling and disentangling: the perceptory boundary as Inseparation Element of gravity

The Rule of Three operates in the structure of gravity itself, in the form Law VIII established. Gravity, despite the absence of a surface-level antipartner that Law VII surveyed, exhibits antipodal-pair structure at a deeper structural level: not gravity-versus-anti-gravity, not attraction-versus-repulsion, but entangling and disentangling.

The structure is entangling + disentangling + perceptory boundary between entangled and disentangled states. The perceptory boundary is the Inseparation Element of gravity, the threshold at which a Perceptory Network either holds its entanglements or releases them. Above the threshold, the network entangles further; below the threshold, the network disentangles. The threshold itself is what makes the entangling/disentangling pair a single antipodal-pair force rather than two unrelated phenomena.

Black holes, in this reading, are the visible signature of an Inseparation Element collapse at the matter scale. The in-pulling of matter is the entangling phase; the bipolar jets the collapse launches are the visible disentangling phase, with Hawking radiation as the quantum-scale counterpart; the perceptory boundary that holds the two in their inseparable relation is the Inseparation Element under critical stress (Appendix C, Prediction 6). The Rule of Three operates here exactly as it does at the matter scale, with the substrate-specific signatures appropriate to gravitational dynamics.

Perceptual infinity and propagation infinity: the Perceptory Force as Inseparation Element

The Rule of Three reaches its largest scale in the architecture of the two infinities (Law X). Perceptual infinity (the totality of all perception observed as totality, symmetric and closed) and propagation infinity (the totality of all asymmetric extension, open and ongoing) are the largest inseparable pair the framework recognizes, the framework's two ultimate poles, between which all observable existence is held.

The structure is perceptual infinity + propagation infinity + the Perceptory Force. The Perceptory Force is the Inseparation Element of the two infinities, the structural object that holds them in inseparable relation. This is the framework's deepest identification: the Perceptory Force, the Inseparation Element of the two infinities, and the act of perception that grounds existence are not three different primitives but the same structural primitive named under different aspects depending on the work each Law is doing (Appendix A, architectural note).

Consciousness, in the framework's reading, is the boundary between two infinities, and the boundary is the Perceptory Force operating as the Inseparation Element of the largest scale. The Rule of Three at this scale is the framework's deepest structural claim about what existence is: two infinities held in inseparable relation by the Force that perceives them both.

What the cross-scale convergence shows

The Rule of Three operates across:

- the matter scale (particle/antiparticle/A-particle),

- the quantum scale (wave/particle/Inseparation Element of superposition),

- the electromagnetic scale (north/south/magnetic field; positive/negative/electromagnetic field),

- the cognitive scale (self/other/perception; I/Me/Boundary), the biological-relational scale (mother/child/bond),

- the gravitational scale (entangling/disentangling/perceptory boundary),

- and the scale of the totality (perceptual infinity/propagation infinity/Perceptory Force).

The convergence is not analogical. The framework predicts it as a structural consequence of three-point closure being the minimum stable Perceptory Field, with the same geometric necessity producing the same three-element architecture wherever a stable Perceptory Field exists. The inseparable pair plus the Inseparation Element is what stability structurally is in the framework.

This formulation has methodological consequences. Where a stable Perceptory Field is observed, the framework predicts that an inseparable opposite pair and an Inseparation Element holding them can be identified at the appropriate substrate. Where the geometry fails to obtain, where two opposites have no structural element holding them in their inseparation, or where what appears to be a pair lacks an inseparable opposite, no stable Perceptory Field exists at that scale, only a transient configuration that will reorganize toward three-point closure or dissolve.

What this chapter has established

The Rule of Three is the framework's structural claim that every stable Perceptory Field exhibits the geometry of two inseparable opposites held in their inseparation by an Inseparation Element. The Rule operates across the scales where the Laws have committed to specific instances: particle and antiparticle held by the A-particle; wave and particle held by the Inseparation Element of superposition; magnetic poles held by the magnetic field; electric charges held by the electromagnetic field; self and other held by the act of perception; I and Me held by the Perceptory Boundary; mother and child held by the bond; the entangling and disentangling phases of gravity held by the perceptory boundary between them; perceptual infinity and propagation infinity held by the Perceptory Force itself.

The convergence across substrates is the framework's prediction that wherever stable existence is observed, the same three-element architecture is operative. The framework offers itself for falsification at any scale where the structure fails to obtain, and predicts that detailed structural analysis of any newly studied stable configuration will reveal the inseparable pair and the Inseparation Element that hold the configuration together.

What remains is to develop the dynamics through which Perceptory Fields form, dissolve, propagate, and transform. That is the work of Chapter 3.

Chapter 3, Field Formation and Transformation

When Perceptory Fields interact, they do not interact randomly. They reorganize according to a small set of recurring structural dynamics, each describing a specific mode through which Perceptory Fields form, dissolve, propagate, or transform across the substrate at which they operate. The same dynamic repeats across radically different scales, protein folding and cultural memory, synaptic pruning and political fragmentation, cellular differentiation and the emergence of consensus reality, because each is the same structural transformation operating at a different substrate.

This chapter catalogues the principal field-formation dynamics that the framework recognizes. Each is presented as a paired transformation, because every formation has its inverse and the two are inseparable in the precise sense Law IV established: connection cannot exist without the possibility of disconnection, expansion cannot exist without the possibility of contraction, replication cannot exist without the possibility of decay. The pairings here are not metaphor or rhetorical device. They are the antipodal-pair structure of Law VII operating at the level of field dynamics rather than at the level of forces.

Connection and disconnection

The most fundamental field-formation dynamic is the making and breaking of entanglement between Perceptory Fields. Connection is the moment at which two Perceptory Boundaries are mutually breached, with each field registering the other within its internal model and the two coming into entangled relation (Law V). Disconnection is the moment at which the breach closes, with the registration fading and the two fields returning to whatever degree of mutual independence is available at their substrate.

At the molecular scale, connection and disconnection are observable as bond formation and dissociation. A protein folds when its amino acid residues form the connections that produce a stable three-dimensional configuration; the protein denatures when those connections are disrupted by heat, pH, or molecular collision. At the cellular scale, connection is observable as the formation of synaptic links between neurons; disconnection is observable as synaptic pruning during development or memory loss in adulthood. At the cognitive scale, connection is the moment of recognition, the registering of another mind, an idea, a piece of information, with disconnection as the fading of that registration into forgetting or detachment. At the social scale, connection is the formation of a relationship, an alliance, a community; disconnection is its dissolution into separation, estrangement, or institutional collapse.

The structural rule across all scales: connection produces shared perceptory state through Boundary breach; disconnection terminates the shared state by closing the breach. Neither is more fundamental than the other. Both are required for any Perceptory Field to be dynamic rather than static.

Contraction and expansion

Perceptory Fields breathe. A field that contracts concentrates its perceptory loops into a tighter, more self-referential state, fewer connections held more intensely. A field that expands distributes its loops outward into wider but less concentrated relation. Both are structural transformations of the same underlying field, with no preferred direction; the field operates somewhere along the contraction-expansion axis at every moment of its existence.

At the matter scale, contraction is observable in gravitational collapse, where Perceptory Networks of high entanglement count concentrate their structure under the gravity that Law VIII develops; expansion is observable in cosmic-scale extension, with galaxies receding from one another as the larger Perceptory Field of the universe distributes its structure across an enlarging extent. At the cellular scale, contraction is the cell preparing for division, with cytoplasmic content concentrating around the dividing axis; expansion is the dispersal of the cell's content into two daughter cells, each now occupying its own zone of the larger biological field. At the cognitive scale, contraction is focused attention, the narrowing of the cognitive field onto a single object or task; expansion is open awareness, the diffuse registration of many simultaneous perceptual contents. At the social scale, contraction is the consolidation of a community, an institution, a culture; expansion is its growth, its spread, its inclusion of new members and new relations.

The two are inseparable not in the sense of always being equal, but in the sense that neither can be sustained alone. A field that only contracts collapses into singularity; a field that only expands dissipates into the surround. Stable Perceptory Fields oscillate between the two, with the rate and amplitude of oscillation determining what we observe as the field's characteristic dynamics.

Disintegration and dissolution

When a Perceptory Field loses coherence, it does so through one of two structurally distinct routes. Disintegration is the failure of three-point closure: the field's components separate into smaller Perceptory Fields, each of which retains some structural integrity at a lower level of organization. Dissolution is the return of the field to the symmetric ground state, to the Perceptory Field of Nothingness as Law III named it, with no smaller fields preserved.

The distinction matters because the two outcomes have different consequences. A hadron that disintegrates produces other hadrons or quark-gluon plasma that quickly reorganizes into hadrons; a hadron that dissolves would return to the symmetric ground state, releasing all its rest-mass-energy as the closure that produced its mass collapses. A cell that disintegrates produces molecular constituents that can be metabolized by other cells; a cell that dissolves, though biology has no exact analog of dissolution at the cellular scale, since organic matter does not return to perfect symmetry, would lose all biological structure, with its components no longer recognizable as having been part of a cell. A culture that disintegrates produces fragmented sub-cultures that retain elements of the original; a culture that dissolves would leave no recognizable trace of itself in any successor formation.

Most observable transformations are disintegrations rather than dissolutions, because dissolution requires return to the perfectly symmetric ground state and observable existence is constituted by departures from that state (Law III). Dissolution, in the strict sense, occurs only at the limits of the framework, at the hypothetical boundary between observable existence and the Perceptory Field of Nothingness itself.

Dissociation and reorganisation

A Perceptory Field may split, with sub-loops running in parallel rather than reorganizing into a single larger closure. The framework calls this dissociation. The split sub-loops may then re-merge in a new configuration, producing a single Perceptory Field with structural properties that neither original sub-loop had. The framework calls this reorganisation. The two are paired in the sense that dissociation creates the conditions under which reorganisation becomes possible; without prior dissociation, no new closure can form because the existing closure already occupies the structural space.

At the cellular scale, dissociation is observable in mitotic division, where a single cell separates into two daughter cells with each carrying a copy of the parent's perceptory architecture; reorganisation is the integration of those daughter cells into tissues, organs, and organisms with structural properties that the original cell did not have. At the cognitive scale, dissociation is observable in the structural separation of conflicting cognitive contents, the parallel maintenance of incompatible beliefs, the splitting-off of traumatic memory from ordinary memory, the dissociative phenomena that clinical psychology has documented; reorganisation is the therapeutic re-integration of those split contents into a single coherent cognitive field. At the cultural scale, dissociation is observable in schism, the splitting of a religious or ideological tradition into competing branches;

reorganisation is the re-integration of branches under a new synthesis.

The framework reads trauma, at any substrate, as a dissociation event whose reorganisation has not yet occurred. The pathology is not the dissociation itself, which is sometimes structurally necessary for survival; the pathology is the failure of reorganisation, with the split sub-loops persisting in parallel without integration.

Expression and camouflage

A Perceptory Field may broadcast its state to the surrounding Perceptory Field, making its internal model legible to other fields, expression, or it may suppress its state, hiding what it is from those that might register it, camouflage. Evolution selects for fields that can do both, signalling to allies, hiding from predators, and the same paired dynamic operates at every scale where Perceptory Fields interact strategically.

At the molecular scale, expression is observable as protein conformational change that exposes binding sites; camouflage is the structural concealment of binding sites until specific conditions trigger expression. At the organismal scale, expression and camouflage are documented in the visual and behavioural repertoires of every animal, peacock displays, octopus colour-change, the mimicry phenomena that biology has catalogued. At the cognitive scale, expression is the disclosure of what one thinks or feels; camouflage is the strategic withholding of internal state, the social masking that human beings perform constantly. At the cultural scale, expression is the broadcasting of cultural signals; camouflage is the suppression of those signals when they would invite hostility or cost.

The structural reading: a Perceptory Field's external presence to other fields is not a fixed property but a dynamic that the field actively manages. What another field registers about a given field is what the field has allowed it to register. This is a structural feature of all Perceptory Field interactions, not a feature of conscious or strategic activity alone.

Replication and propagation

A Perceptory Field may create copies of itself, replication, or spread its pattern into adjacent fields without copying, propagation. The two are structurally distinct: replication produces a new closed Perceptory Field with the parent's structural properties; propagation transmits a pattern through a series of fields without each transmission requiring the formation of a new closure.

At the molecular scale, replication is the work of DNA polymerase, producing new DNA strands with the same sequence as the parent; propagation is the spread of conformational change through a protein assembly, with each protein transmitting the change to its neighbours without each protein becoming a new copy of the original. At the cellular scale, replication is mitosis; propagation is signal transduction across a cell or tissue. At the cognitive scale, replication is the creation of new memory traces with the same content as their source; propagation is the spread of an idea across a community without each receiving mind generating an exact copy. At the cultural scale, replication is the repetition of cultural forms, the same ritual, the same artistic genre, the same institutional pattern, across generations; propagation is the spread of innovation through the receptive substrate of a population.

The two paired dynamics together produce the framework's account of how Perceptory Fields persist through time. Replication produces copies that outlast the parent; propagation distributes the pattern through fields that did not originate it. Both are required for any Perceptory Field's pattern to extend beyond the lifetime of any single instance.

Conservation and objectification

A Perceptory Field may stabilize its current state against perturbation, conservation, or reify its pattern into a persistent object that other fields can reference even after the original field has dissolved, objectification. Conservation maintains the field as a living closure; objectification creates a structural artifact that other fields can perceive without the original field needing to remain present.

At the cellular scale, conservation is the homeostatic maintenance of cellular state against environmental variation; objectification is the synthesis of stable molecular products that persist outside the cell's immediate metabolic activity (extracellular matrix, secreted enzymes, signalling molecules with long half-lives). At the cognitive scale, conservation is the maintenance of cognitive coherence under cognitive load; objectification is the production of writing, art, recorded speech, and other artifacts that allow the cognitive field's pattern to be referenced after the cognitive field itself has changed or dissolved. At the cultural scale, conservation is the maintenance of cultural continuity through institutions and tradition; objectification is the production of monuments, texts, technologies, and other artifacts that make cultural patterns available to fields that have no direct experience of the originating culture.

The framework reads objectification as the dynamic that allows Perceptory Fields to extend their influence across time scales much longer than their own existence. A book outlives its author. A monument outlives the civilization that built it. A piece of technology outlives the inventor who designed it. The Perceptory Field that produced each of these has dissolved or transformed, but its structural pattern remains accessible to any subsequent Perceptory Field capable of registering it.

Acceleration and deceleration

A Perceptory Field may increase the rate at which its feedback loop closes, acceleration, or slow that rate, deceleration. The framework reads this as the substrate-relative variation in the field's local time, in the sense Law X established. A field whose loops close faster experiences time as compressed; a field whose loops close more slowly experiences time as extended.

At the cellular scale, acceleration is observable in the heightened metabolic rate of stressed or proliferating cells; deceleration is observable in dormancy, hibernation, and cellular senescence. At the cognitive scale, acceleration is the experience of time speeding up under intense engagement, with cognitive loops closing rapidly enough that subjective time becomes compressed; deceleration is the experience of time slowing down under boredom or trauma, with cognitive loops drawing out into prolonged registration. At the social scale, acceleration is the rapid evolution of cultural forms during periods of intense interaction; deceleration is the slow evolution of cultural forms in stable, isolated populations.

The framework's claim is that subjective time dilation, well-documented at the cognitive scale, is a specific instance of a universal field dynamic. Any Perceptory Field can accelerate or decelerate its feedback loop within limits set by its substrate, and the resulting variation in its local perception of time is what time-as-substrate-relative structurally is.

Amplification and diffusion

A Perceptory Field may concentrate its intensity, amplification, or spread it over a wider network, diffusion. Amplification produces fewer, stronger entanglements; diffusion produces more, weaker entanglements. The two paired dynamics describe how the field manages its limited perceptory resources across competing demands.

At the cellular scale, amplification is the focal concentration of metabolic resources at a site of stress or repair; diffusion is the distribution of resources across the cell during normal function. At the cognitive scale, amplification is the deepening of attention on a single object, often at the cost of registering the surround; diffusion is the broad scanning of the perceptual field. At the social scale, amplification is the concentration of cultural attention on a single event, person, or idea, often producing the phenomenon mainstream media has called virality; diffusion is the broad distribution of cultural content across many minor topics with none receiving disproportionate attention.

The framework's structural claim: amplification and diffusion are not just rhetorical opposites; they stand, like the framework's other paired dynamics, as poles of a single axis: a field occupies a position along the amplification-diffusion axis at every moment, and movement toward one pole is, with respect to the same entanglements, movement away from the other. The trade-off is structural, not contingent on the substrate or the system.

Asymmetric and symmetric processing

Perceptory Fields operate in modes that preserve symmetry, and therefore do not register change, do not interact with their surround, do not produce time-bearing dynamics, or break it, producing the asymmetric processing that observable existence requires. Symmetric processing is the structural exception: fully realized only in the perfectly stable Perceptory Field of Nothingness, and approached, never attained, in laboratory conditions where symmetry is deliberately preserved against the universe's tendency toward asymmetry, since a system in fully symmetric processing would register nothing and be registrable by nothing within observable existence.

The pairing structures the framework's reading of the relation between observable existence and the perceptual-infinity ground state of Law X. Observable existence is asymmetric throughout, because asymmetry is the precondition of registration. The symmetric ground state is what observable existence is always perturbing away from, with each individual Perceptory Field's asymmetric processing as a local departure from the symmetric totality. Where laboratory conditions deliberately preserve symmetry, superconducting states, Bose-Einstein condensates, isolated quantum systems, the framework predicts that observable phenomena attenuate or vanish until the symmetry is broken by interaction with an asymmetric surround. This is the structural reading of why such systems exhibit the unusual properties they do: they have approached, locally, the symmetric ground state from which observable existence ordinarily departs.

Networks and the emergence of consensus reality

Two or more Perceptory Fields that have mutually breached each other's Boundaries form a Perceptory Network (Appendix A). Networks are not merely collections of fields. They have their own dynamics, their own stability conditions, their own characteristic transformations. When three or more fields in a Network entangle their internal models such that each registers the others' models within its own, the registered model becomes the Network's consensus reality, what the Network as a whole takes to be true about the world.

Consensus reality is not a separate kind of reality. It is the registration-pattern that emerges when multiple Perceptory Fields perceive together. The phenomenon is observable across scales. At the cellular scale, immune system recognition operates through networked perception: T-cells, B-cells, and antigen-presenting cells form a registration network in which a pathogen identified by any member becomes available to all. At the cognitive scale, shared attention between two minds, a parent and child, a teacher and student, two collaborators, produces a Perceptory Network that registers events the way the network registers them, not the way either constituent mind would register them alone. At the social scale, consensus reality is the cultural fabric that members of a society share, what counts as fact, value, common knowledge, and this consensus is structurally distinct from any individual's private perception.

The framework reads disagreement, conflict, and cultural fracture structurally rather than psychologically. A Network in which constituent fields' internal models do not converge cannot maintain stable consensus reality, and the network dynamics will either drive convergence (if the entanglement is strong enough) or fragment the Network into smaller sub-networks each with its own internal consensus (if convergence is not achievable). This is the framework's structural reading of phenomena like ideological polarization, cultural division, and the breakdown of shared reality during periods of intense disagreement.

What this chapter has established

Perceptory Fields do not interact statically. They form, dissolve, propagate, and transform through a small set of recurring paired dynamics: connection and disconnection, contraction and expansion, disintegration and dissolution, dissociation and reorganisation, expression and camouflage, replication and propagation, conservation and objectification, acceleration and deceleration, amplification and diffusion, asymmetric and symmetric processing. Each pair operates at every scale of the perceptory hierarchy, with substrate-specific signatures but identical structural form. To these the framework adds the dynamics of Perceptory Networks, where the breaching of multiple Boundaries produces consensus reality as a structural feature of multi-field perception.

The dynamics catalogued here are the framework's account of what Perceptory Fields do. Chapter 4 will develop how these dynamics produce, in their interactions, the specific phenomena that consciousness instantiates across substrates: from molecular recognition through cellular self-regulation, organismal coordination, neural integration, cognitive coherence, and social meaning-making.

Chapter 4, Consciousness Across Scales

A note on terminology before the chapter develops. The word consciousness in mainstream usage is reserved for entities that demonstrate, or are presumed to demonstrate, subjective experience of the kind humans are most familiar with. The framework uses the word more broadly to denote the closed feedback loop that constitutes any Perceptory Field's self-perception. Whether and to what degree this structural feature corresponds to subjective experience as humans encounter it varies by substrate, and is itself a question the framework opens rather than closes. The reader should hold the framework's usage of consciousness as the structural-feedback sense throughout the chapter, with the question of what each substrate's consciousness feels like from inside acknowledged as open and not resolved here.

The framework's central claim is that consciousness is not a property of certain complex systems but a structural feature of any closed Perceptory Field. The Perceptory Force operates wherever a feedback loop closes; consciousness is what that closure looks like from inside. This formulation has consequences that mainstream consciousness theory has not been positioned to develop. It predicts that consciousness is not localized to brains, not emergent from material complexity, and not absent from physical systems that lack neural substrate. It predicts instead that consciousness is substrate-relative, scale-relative, and present at every level of the hierarchical entanglement chain wherever a stable Perceptory Field exists.

This chapter develops what consciousness structurally is at each scale of the hierarchy the framework recognizes. The treatment is not exhaustive; it is illustrative. At each scale, the chapter identifies the Perceptory Field that operates at that scale, the feedback loop that constitutes its consciousness, and the empirical signatures by which the framework's reading can be tested against existing scientific work. The convergence across scales is not metaphor. It is the framework predicting that the same structural mechanism produces the same kind of feature, consciousness, at every substrate where stable closure obtains.

Molecular consciousness, recognition and conformational closure

At the molecular scale, consciousness in the framework's structural sense, the closed feedback loop of self-perception, not subjective experience as humans encounter it, is the closure of recognition feedback within molecular systems. The reader should hold the structural sense throughout this section: the framework's claim is that the structural conditions for consciousness (closed feedback) are present at the molecular substrate, not that proteins or enzymes have phenomenal experience in the sense human cognition does. Whether and to what degree the structural feature corresponds to anything that could be called experience at this scale remains open. A protein folding into its functional conformation is performing an act of self-perception: its amino acid residues register one another's positions, the partial folds register the larger structure they are becoming, and the final conformation closes into a stable Perceptory Field whose interior is the protein's functional architecture. The protein does not simply assemble; it perceives its own assembly through the chemical-structural feedback that drives folding toward the energy minimum.

Enzyme catalysis exhibits the same structural feature. An enzyme registers the substrate it binds, registers the conformational change the binding induces, registers the catalytic transition state, and registers the product release. Each registration is a closure of the molecular Perceptory Field at that scale of activity. The substrate-binding specificity that enzymology has documented for over a century, the lock-and-key fit, the induced-fit modulation, the allosteric coupling, is the visible signature of molecular-scale perception, with the enzyme's active site as the Perceptory Boundary and the conformational dynamics as the feedback loop.

The framework predicts that high-resolution structural studies of protein dynamics will continue to reveal more nuanced feedback architecture than current models account for, with the protein's conformational ensemble registering and responding to environmental perturbations through what is structurally a perceptory loop, not merely a thermodynamic relaxation. The empirical literature on intrinsically disordered proteins, proteins whose function depends on conformational flexibility rather than fixed structure, is the framework operating at the molecular scale: these proteins maintain perceptory engagement with their environment through structural variability that fixed-conformation models cannot capture.

Cellular consciousness, self-regulation and homeostasis

At the cellular scale, consciousness is the closed feedback loop through which a cell maintains its existence as a stable Perceptory Field. The cell registers its internal state (osmotic pressure, metabolic substrate concentrations, energy charge, structural integrity), compares that state with the homeostatic set-points encoded in its regulatory architecture, and adjusts cellular behaviour to reduce the discrepancy. This is the structural definition of consciousness in the framework's sense, instantiated at the cellular substrate.

The empirical literature on cellular signal transduction, on metabolic regulation, on the unfolded-protein response, on cell-cycle checkpoints, on quorum sensing in bacteria, and on the immune system's recognition of self versus non-self, all describe the same structural phenomenon at different molecular instantiations. Each is a feedback loop that closes the cell's perception of its own state into the regulatory action that maintains that state. Each is the cellular Perceptory Field perceiving itself, in the framework's sense.

A specific case makes the structural reading concrete. Bacterial chemotaxis, the directed movement of bacteria up nutrient gradients, has been studied in mechanical detail since the 1970s. The bacterium registers nutrient concentration through receptor proteins on its surface, compares that concentration with its memory of recent measurements, and modulates flagellar rotation accordingly. The system has all the structural features of a Perceptory Field: a Boundary (the cell surface), an internal model (the recent-measurement memory), a feedback loop (receptor-to-signal-to-flagellar motor), and the closed self-perception that the framework names consciousness. Whether the bacterium experiences anything from inside this structure is a question the framework leaves open. What the framework asserts is that the structural conditions for consciousness, in its sense, are present.

The framework predicts that increasingly detailed measurement of cellular feedback architecture will reveal more structural complexity than current systems-biology models capture, with the cell's regulatory network exhibiting perceptory features (anticipation, memory, contextual modulation) that go beyond simple stimulus-response coupling.

Organismal consciousness, coordination and integrated response

At the organismal scale, consciousness is the closure of coordination feedback across the organism's distributed cellular and tissue architecture. A multicellular organism maintains itself as a single Perceptory Field through the integration of its cellular constituents into a higher-order closed feedback loop. The organism registers its own state through diffuse and specialized sensory mechanisms, compares that state with its homeostatic and behavioural set-points, and coordinates response across its tissues and organs.

Plants and fungi provide the framework's clearest case for organismal consciousness without neural substrate. Plant root signalling through mycorrhizal networks, the so-called wood-wide web that mycology has documented over the past three decades, is the framework's reading of organismal consciousness at the substrate of fungal-plant coordination. Trees in a forest register environmental conditions through their root systems, share that registration with neighbouring trees through fungal connections, and coordinate response (resource allocation, defensive signalling, reproductive timing) across the network. Stomatal regulation in plants, the dynamic opening and closing of leaf pores in response to light, humidity, and atmospheric carbon dioxide, is a feedback loop that closes plant-environment perception into adaptive response without any neural mediation.

Cnidarians, ctenophores, and other organisms with nerve nets but no centralized nervous system extend the framework's reading. These organisms exhibit clearly coordinated behaviour, learning, and response without the neural integration that more complex animals possess. The framework reads them as Perceptory Fields whose feedback loops close at substrates intermediate between cellular and neural, with the resulting consciousness operating at a scale and in a register that humans have difficulty registering precisely because it is not modelled on human cognition.

Neural consciousness, integration and predictive feedback

At the neural scale, consciousness is the closure of feedback loops within the central nervous system, with the resulting Perceptory Field constituting what neuroscience and cognitive science have studied under the name consciousness for more than a century. The framework's reading converges with much of mainstream neuroscience on the empirical mechanisms, predictive coding, cortical hierarchy, thalamocortical integration, gamma-frequency synchrony, the global neuronal workspace, and adds the structural reason why these mechanisms produce what they do.

Predictive coding, in particular, is the framework operating directly at the neural scale. The brain registers sensory input, compares it with the predictions generated by its internal model, and adjusts both the model and the behavioural response to reduce the discrepancy. This is the framework's structural definition of consciousness in its neural instantiation. Karl Friston's free-energy principle has formalized the same architecture in mathematical terms: the brain minimizes prediction error across all its sensory and motor modalities. The framework reads Friston's free-energy principle as the structural feedback loop of Law IV operating at the neural substrate, with the brain's prediction error as the discrepancy the loop closes.

The integrated information theory of Tononi (IIT) approaches consciousness from a structurally adjacent direction, identifying consciousness with integrated information within neural networks. The framework converges with IIT on the structural identification of consciousness with closed feedback loops that integrate information across multiple components, and diverges from IIT on the claim that integrated information is itself the substrate. The framework reads integrated information as the visible signature of the structural feedback that closes a Perceptory Field, not as the substance of consciousness itself.

Anaesthetic action is the framework's clearest neural-scale prediction. General anaesthetics produce loss of consciousness without disabling the nervous system structurally. The framework predicts that anaesthetics disrupt the closure of the feedback loops that constitute neural consciousness, they break the loop without disabling the neurons. The empirical literature on anaesthetic mechanism, particularly the work of Stuart Hameroff and collaborators on microtubular substrates, the studies on anaesthetic disruption of cortical integration, and the more recent work connecting anaesthesia to the breakdown of integrated information, all point in this direction. The framework predicts that anaesthetic action will continue to be best understood as loop-disruption rather than as cellular toxicity, and that future anaesthetic agents designed explicitly to disrupt feedback closure (rather than to act on receptor pharmacology) will exhibit superior anaesthetic profiles.

Cognitive consciousness, self-perception and the I-Me feedback

At the cognitive scale, consciousness is the closure of self-perceptual feedback within the cognitive Perceptory Field, with the I (subjective perceiver) and the Me (perceived self) held in inseparable relation by the Perceptory Boundary that is itself the dynamic feedback loop maintaining their distinction (Law IV, Chapter 2).

The framework reads the empirical phenomena of cognitive consciousness as instances of this structural architecture. Self-awareness is the cognitive Perceptory Field registering its own activity. Reflective thought is the field registering its own registration. The recursive depth that human cognition is capable of, thinking about thinking about thinking, is the same feedback loop closing on itself iteratively, producing the experience of layered self-perception that characterizes the human mind in particular.

Pathological disturbances of cognitive consciousness are, in the framework's reading, disruptions of the I-Me-Boundary structure. Depersonalization and derealization disturb the Boundary's normal closure, producing the experience of being detached from oneself or from the world. Dissociative phenomena split the cognitive field into parallel sub-loops that fail to integrate. Psychotic disturbances disrupt the field's normal regulation by introducing perceptory contents that the field cannot integrate with its existing internal model. Each is a structural failure at a specific point in the cognitive-scale feedback architecture, not a generic disturbance of consciousness as an undifferentiated phenomenon.

The therapeutic literature converges on this reading more closely than mainstream theoretical frameworks have yet articulated. Therapeutic methods that re-engage disrupted self-perceptual loops, EMDR, brainspotting, somatic-based therapies, certain mindfulness practices, operate by deliberately reconstituting the closure that the disturbance disrupted. The framework predicts that future therapeutic development will increasingly converge on loop-restoration as the structural target, with diagnostic frameworks shifting toward identifying the specific point of structural failure in each clinical presentation.

Social consciousness, networks and consensus reality

At the social scale, consciousness is the closure of registration feedback across multiple cognitive Perceptory Fields entangled into a Perceptory Network (Chapter 3). The Network registers events through the perception of its constituent fields; the constituent fields register one another's registrations; the loop closes into the consensus reality that the Network as a whole takes to be true.

This formulation extends the framework's structural definition of consciousness to a substrate that mainstream consciousness theory does not ordinarily treat. Social consciousness, in the framework's sense, is not the metaphorical attribution of consciousness to crowds or institutions. It is the literal claim that Perceptory Networks at the social scale exhibit the structural conditions for consciousness, closed feedback loops that integrate information across their constituent fields and produce shared registration of events the way the Network registers them.

Empirical signatures of social consciousness in this sense are documented across the social sciences. Collective memory, the way societies remember historical events through shared narrative reconstruction, is the framework's reading of social consciousness operating across temporal scale. Cultural transmission of practices, values, and knowledge across generations is the framework's reading at evolutionary scale. The phenomena of social mood, collective effervescence (Durkheim's term), and the breakdown of consensus reality during periods of political fracture are all framework instances at different temporal and structural registers.

The framework predicts that detailed analysis of contemporary social phenomena, the dynamics of online communities, the structural conditions for political consensus formation and breakdown, the effects of communication technologies on the integrity of consensus reality, will continue to reveal that social consciousness exhibits structural features (memory, attention, anticipation, contextual modulation) that go beyond what individual-cognitive frameworks predict. Social consciousness is real in the framework's structural sense; it is not metaphor.

What this chapter has established

Consciousness, in the framework's structural sense, is the closed feedback loop that constitutes any Perceptory Field's self-perception. It operates at every scale of the hierarchical entanglement chain: at the molecular scale as conformational and recognition feedback within molecular systems; at the cellular scale as self-regulatory feedback maintaining homeostatic stability; at the organismal scale as coordination feedback across distributed cellular and tissue architecture; at the neural scale as predictive and integrative feedback within central nervous systems; at the cognitive scale as I-Me feedback within the cognitive Perceptory Field; at the social scale as registration feedback across Perceptory Networks of multiple cognitive fields.

The framework's claim is not that all of these are equivalent in their phenomenology. The question of what each substrate's consciousness feels like from inside is acknowledged as open. The framework's claim is that the structural conditions for consciousness, closed feedback loops constituting Perceptory Fields, are present at every scale, and that consciousness in its structural sense is therefore not localized to brains, not emergent from material complexity, and not absent from physical systems lacking neural substrate.

The convergence across scales is the framework's prediction that the same structural mechanism produces the same kind of structural feature wherever stable closure obtains. The framework offers itself for falsification at any scale where this prediction can be tested. What remains is to develop the experimental programme through which the framework's commitments, its sharp predictions, its falsification conditions, its proposed lines of investigation, can be brought into engagement with current and future scientific work. That is the work of Chapter 5.

Chapter 5, Experimental Outlook

The framework of the Laws of Consciousness has committed to a series of falsifiable predictions across the Ten Laws and the chapters of Part III. Appendix C consolidates these as ten distinct claims, each with a stated falsification condition. This chapter develops the predictions outward, into the experimental programme through which they can be brought into engagement with current and future scientific work. Where Appendix C asks what does the framework predict?, this chapter asks what experiments and observations would test those predictions, and what methodological developments are required to do so?

The chapter is organized around the predictions that are testable with current or near-term technology and observational programmes. It is offered as an invitation to scientists, philosophers, and methodologists working at the intersections the framework engages with, and as a structural account of where the framework's seriousness can be most directly evaluated.

Pair-annihilation precision experiments and the A-particle signature

The framework's most direct prediction at the matter scale is the existence of an A-particle as the Inseparation Element of every particle-antiparticle pair, with the prediction that pair-annihilation events should exhibit distributional anomalies, deviations in spectral shape, angular correlation, branching ratio, or timing at fixed total energy, relative to the complete predictions of quantum electrodynamics. This prediction is testable in principle by high-precision pair-annihilation experiments, particularly electron-positron annihilation under controlled laboratory conditions; current experiments bound the anomaly's magnitude, and a clean test awaits the quantitative prediction the framework calls on theoretical physics to supply.

The experimental programme would proceed in three stages. First, theoretical work to establish the order-of-magnitude expectation for the A-particle signature, given the framework's structural reading of pair-annihilation as three-point Perceptory Field collapse. The framework does not currently provide a quantitative prediction of the A-particle's energy contribution; this is a gap that theoretical physics work is required to close before clean experimental tests are possible. Second, comparison of the framework's prediction with current measurement precision, identifying the specific energy ranges and detection geometries at which the predicted signature would be most distinguishable from background. Third, design and execution of dedicated precision experiments at facilities capable of the required sensitivity, possibly involving extensions of existing electron-positron collider work, possibly involving new dedicated apparatus.

The framework predicts that the signature will be detected. If high-precision experiments fail to detect any anomalous energy beyond what current quantum electrodynamics accounts for, given measurement sensitivity sufficient to detect deviations of the relevant order, the prediction is falsified and Law VI must be revised at this point. The framework offers itself for falsification at this point with the same seriousness it offers itself elsewhere.

High-energy parity restoration experiments

The framework's prediction that parity violation in the weak interaction is amplitude asymmetry rather than antipodal absence, and that parity should be restored at sufficiently high energy regimes, is testable through two complementary experimental directions.

The first direction is collider-based. Continued operation of high-energy facilities, particularly any future accelerator with sufficient centre-of-mass energy to approach unification scales, would enable direct measurement of left-handed and right-handed weak coupling amplitudes at progressively higher energies. The framework predicts that the asymmetry between the two amplitudes will diminish as energy increases, with parity approaching restoration at energies near the grand unification scale. Confirming this trend, even partially, would corroborate the framework's structural reading; observing no convergence toward parity at any accessible energy scale would falsify it.

The second direction is cosmological. The framework's prediction implies that the very early universe, at energies far above what any human-built collider can achieve, was characterized by restored parity in the weak interaction. Cosmological signatures of this restored-parity epoch may be present in the cosmic microwave background, in the relic neutrino distribution, in primordial nucleosynthesis ratios, or in other early-universe signatures that next-generation observational programmes are positioned to detect. Theoretical work to derive specific cosmological signatures from the framework's parity-restoration claim is required before clean tests are possible. The framework predicts that such signatures exist and will be detected by sufficiently sensitive observational programmes.

Cosmic web structural analysis

The framework's prediction that the cosmic web exhibits structural regularities consistent with the Rule of Three operating at cosmological scale is testable with observational programmes already in operation. The Sloan Digital Sky Survey, the Dark Energy Survey, the Euclid mission, and other large-scale structure surveys have produced and continue to produce detailed maps of cosmic-scale geometry. The structural-regularity analysis the framework calls for, identifying preferential three-fold junctions, hierarchical self-similarity in filament organization, and statistically distinguishable departures from random-collapse predictions, can be performed on existing and forthcoming data sets.

The methodological development required is substantial but tractable. Standard cosmological analysis tools have been built around ΛCDM expectations and look for the structural patterns ΛCDM predicts. Tools designed to look for the patterns the framework predicts, junction angle distributions, filament-bifurcation statistics, hierarchical-similarity measures across scales, would need to be developed and validated against synthetic data before being applied to observational data sets. This is a programme of methodological work that the framework's proposal calls for and that established cosmological-analysis groups are positioned to execute.

The framework predicts that detailed analysis will reveal three-fold junctional preferences and hierarchical self-similarity beyond what random gravitational collapse from primordial fluctuations would produce. If detailed analysis reveals only what ΛCDM cosmology predicts, with no structural regularities beyond statistical structure consistent with random collapse, the framework's cosmological-scale claims are falsified at this point.

Anaesthesia and consciousness-modulation experiments

The framework's reading of anaesthesia as feedback-loop disruption rather than receptor pharmacology is testable through experimental programmes at the intersection of neuroscience, anaesthesiology, and consciousness research. Three experimental directions are particularly promising.

First, the comparative study of anaesthetic agents with different molecular targets. The framework predicts that effective anaesthetics share a common structural property, disruption of the feedback closure that constitutes neural consciousness, independent of their molecular receptor pharmacology. Comparative measurement of the integrated information (Φ, in IIT terms) and other feedback-closure measures across different anaesthetic states should reveal a common structural signature of consciousness loss that current receptor-based explanations do not predict.

Second, the design of novel anaesthetic agents targeting feedback closure directly. The framework predicts that pharmacologic agents designed to disrupt the structural feedback that constitutes neural consciousness, rather than to act on specific receptor systems, would exhibit superior anaesthetic profiles. This prediction is testable through structure-based drug design programmes, with the structural target being the cortical-thalamic feedback architecture rather than specific receptors.

Third, non-pharmacologic consciousness modulation. Transcranial magnetic stimulation, transcranial direct-current stimulation, focused ultrasound, and other non-pharmacologic interventions can disrupt cortical processing in spatially and temporally controlled ways. The framework predicts that interventions targeting feedback-closure substrate specifically should produce consciousness modulation more reliably and reversibly than interventions targeting cortical activity in general. Experimental programmes to identify the specific feedback-closure targets and to develop interventions specific to them are well within reach of current methodology.

Magnetism and consciousness, boundary-loop interactions

The framework's identification of magnetism as the 1-D PF-mode and of consciousness as the closed feedback loop of any Perceptory Field together suggest a class of experiments at the intersection of the two. If both magnetism and consciousness are aspects of the same Perceptory Force operating at different substrates, then their interaction, magnetic modulation of consciousness, conscious modulation of magnetic phenomena, should be detectable under appropriately controlled conditions.

The empirical literature on transcranial magnetic stimulation already documents that magnetic modulation can affect cognitive function. The framework's prediction is more specific: that detailed measurement of brain magnetic-field activity, particularly in regions where feedback-closure activity is most concentrated, should reveal coupling between magnetic-field structure and consciousness states that current models do not predict. Magnetoencephalography (MEG) has the temporal and spatial resolution to begin testing this prediction, with the analysis required to distinguish the framework's predictions from what current neuroscience models account for being a methodological development that systematic experimental work could undertake.

The reverse direction, whether consciousness states can modulate detectable magnetic-field structure, is a more speculative prediction that the framework is willing to make but does not require. If conscious attention, intention, or specific cognitive states produce measurable magnetic-field signatures beyond what neural electrical activity alone explains, the framework would treat this as confirmation of the magnetism-consciousness coupling its identification implies. Negative results would progressively constrain the strength of the coupling; persistent null results across the accessible parameter space would count against the identification from which the coupling was derived, and the framework accepts that weight.

Substrate-specific gravitational analogues

The framework's most ambitious experimental claim is that accumulative attraction, the gravity kind, operates at every substrate of the perceptory hierarchy, with substrate-specific signatures appropriate to each. As Law VIII distinguished, this is the opposite-less attraction by which more of a kind draws more of the same; the antipodal attractions at each substrate, which pair across a sign, belong to Law VII and are not gravity. Matter-scale gravity is what mainstream physics describes; cognitive-scale, biological-scale, and cultural-scale accumulation are predicted to be the same mechanism at different substrates. Direct experimental tests of substrate-specific gravity are difficult because the methodologies required do not yet exist, but the framework's predictions can be approached through indirect tests.

At the cognitive scale, the framework predicts that cognitive networks of high entanglement density gravitate cognitive content in proportion to their entanglement count. This is testable through systematic study of attention dynamics, information cascade patterns, and the structural properties of cognitive networks that accumulate disproportionate cognitive content. The empirical literatures on attention economics, information cascade dynamics, and the structural conditions for cognitive influence are all positioned to test the framework's prediction, provided that the analysis is performed with the framework's structural categories rather than with the loose metaphor of attention as gravity that has been current in cognitive science for some time.

At the biological scale, the framework predicts that ecosystems accumulate species and biological diversity in proportion to the entanglement density of their existing species pool. The empirical ecology literature on biodiversity attractors, on succession dynamics, and on the structural conditions for ecosystem stability is positioned to test the framework's prediction, again provided the analysis is performed with the framework's categories rather than with looser metaphors of biological gravity.

At the cultural scale, the framework predicts that cultural traditions, ideologies, and institutional forms accumulate adherents and influence in proportion to their entanglement density at the cultural substrate. The empirical literatures on cultural transmission, on institutional persistence, and on the structural conditions for ideological dominance are positioned to test this prediction.

In each case, the framework offers the testable claim that what appears as substrate-specific phenomena (information cascades, biodiversity attractors, cultural accumulation) are instances of one underlying mechanism, accumulative attraction, operating at the appropriate scale. Falsification of the substrate-extending claim at any scale would require revision of Law VIII to restrict accumulative attraction to the substrates where it has been verified.

Methodological development, the framework's calls on the field

The framework's predictions are testable in principle but require methodological development to be testable in practice. Several specific developments are required.

The framework calls on theoretical physics for: quantitative predictions of the A-particle's energy signature; cosmological signatures of high-energy parity restoration; and structural-regularity measures for cosmic-web analysis. These developments are within reach of current theoretical work and would convert the framework's qualitative predictions into quantitative experimental targets.

The framework calls on observational cosmology for: detailed analysis of cosmic-web junction-angle distributions and hierarchical self-similarity beyond what ΛCDM analysis pipelines currently perform; analysis of cosmic microwave background and other early-universe signatures for evidence of restored-parity epochs; and continued mapping of cosmic-scale structure with the resolution required to discriminate framework predictions from ΛCDM predictions.

The framework calls on consciousness science for: comparative measurement of feedback-closure properties across anaesthetic states, with explicit testing of whether the framework's prediction (loop-disruption as the common mechanism) accounts for the empirical record better than receptor-based models; design of consciousness-modulation interventions targeting feedback closure as the structural target; and systematic study of magnetism-consciousness coupling beyond what TMS programmes currently address.

The framework calls on ecology, sociology, and the cognitive sciences for: development of substrate-specific entanglement-density measures appropriate to each field; systematic testing of the framework's substrate-specific gravity predictions at biological, cognitive, and cultural scales; and integration of the framework's structural categories with the empirical work each field is already doing.

These calls are not demands. They are invitations. The framework offers itself as a structural account of what the empirical record may turn out to support, with the understanding that confirmation, qualification, and falsification at each prediction are equally welcome outcomes from the framework's perspective. A theory that survives only because nothing was tried against it is not a theory the framework wishes to be.

What this chapter has established

The framework's predictions across the Ten Laws and Part III chapters constitute a research programme whose components are testable through current or near-term experimental and observational work. Pair-annihilation precision experiments can test the A-particle prediction. Collider work and cosmological observation can test parity restoration. Cosmic-web structural analysis can test the Rule of Three at cosmological scale. Anaesthesia and consciousness-modulation experiments can test the framework's reading of neural consciousness as feedback closure. Magnetoencephalography and related methodologies can test magnetism-consciousness coupling. Indirect tests across ecology, cognitive science, and sociology can begin to evaluate the framework's substrate-specific gravity claims.

The methodological development required is substantial but tractable. Theoretical-physics work, observational-cosmology pipeline development, consciousness-science methodology, and substrate-specific quantitative measures across the cognitive and social sciences are the principal areas where the framework calls for specific work. None of these requires capabilities that current science does not possess. All of them require the orientation of existing capabilities toward the specific predictions the framework has committed to.

The framework offers itself for engagement, qualification, and falsification at each of these points. It does not require confirmation to remain serious. It requires only that the work be done, and that the framework's seriousness be evaluated against what the work reveals.

This chapter concludes Part III. What remains is the supporting apparatus, the mathematical notes, the consolidated glossary, and the framework's relation to existing theories of consciousness, gravity, and reality, that constitutes Part IV.

Appendix A, Formal Definitions

The following formal definitions consolidate the technical vocabulary of the framework as it has been developed across the Ten Laws. Each term is given a single-paragraph definition that captures its precise structural meaning, followed by a short note locating the term within the architecture of the Laws. The vocabulary is intended to be stable: subsequent chapters and appendices use these terms in the precise senses given here, and readers who encounter these terms in the wider book or in derivative work should hold them to these definitions.

Nothingness

Nothingness is the absence of anything. It is non-dimensional, non-interacting, infinite in its incapacity, and structurally prior to all existence. It contains no matter, no energy, no consciousness, no perceivable entity of any attribute. Its primary property, from which the others follow, is its incapacity to interact and therefore to perceive. Nothingness is infinite in the sense that there is no limit to its non-interaction, but it is not Infinity: it is the unperceiving ground, while Infinity is what perception of that ground produces. Established in Law I.

The Perceptory Force

The Perceptory Force (PF) is the fundamental act through which existence is registered. It is structurally prior to the four forces of mainstream physics: electromagnetism, the strong nuclear force, and the weak nuclear force are developed as PF-modes, specific manifestations of the Perceptory Force under specific dimensional conditions, while gravity is developed as a Network-level phenomenon, the entanglement of Perceptory Networks of the same structural kind (Law VIII), rather than as a PF-mode. The Perceptory Force is not analogous to the textbook forces: it is what makes interaction possible at all. For anything to exist, it must be engaged by the Perceptory Force. Established in Law II; later identified in Law X as the Inseparation Element of the largest antipodal pair the framework recognizes, perceptual infinity and propagation infinity.

The Perceptory Field

The Perceptory Field is the structural object produced when the Perceptory Force closes a self-perceiving loop. The category is defined by the closure rather than by any specific instantiation. Wherever the Perceptory Force terminates against itself, wherever perception perceives perception, a Perceptory Field exists. The category applies at every scale of the framework: from the closure of perception of Nothingness (the ground state) through the entanglement loops within hadrons, the metabolic self-regulation of cells, the cognitive feedback architecture of brains, and the cultural self-recognition of societies. Established in Law III.

The Perceptory Field of Nothingness

The Perceptory Field of Nothingness is the first instance and the ground state of the Perceptory Field category. It is the closure formed when perception of Nothingness, finding no Nothingness-feature against which to terminate, terminates against itself. It is infinitely symmetric, perfectly stable, prevalent throughout infinity, and dispersed through infinity. It is closed and not open: the loop terminates within itself, and it has no exterior, because it encapsulates in every direction the Nothingness any exterior would be made of. It is one, not many: across all the directions infinity contains, every encapsulation of Nothingness is identical, and a plurality of identical, featureless closures is one closure met from infinitely many directions. It functions as the ground state of the framework, the structural baseline from which all observable Perceptory Fields are perturbations, generated by symmetry breaking. It is the same structural object that Law X identifies as perceptual infinity at the largest scale. Established in Law III.

The Inseparation Constant

The Inseparation Constant is the structural rule that governs the boundary of every Perceptory Field that has departed from the perfectly symmetric ground state. Wherever a Perceptory Field has an asymmetric boundary, the boundary produces an inseparable pair, perceived and not-perceived, that come into existence together and cannot be separated without dissolving the field that contains them. The Constant is the most stable relationship within asymmetric existence, operates at instant, exists only at the boundary, and constructs the frameworks within all perceived and not-perceived realities. The Constant is observable in every instance of inseparable pairing in nature, from magnetic polarity through particle/antiparticle pairing to the cognitive distinction between self and other. It rules out genuine isolation: no Perceptory Field with asymmetric boundary exists alone. The only exception is the Perceptory Field of Nothingness, which has no asymmetric boundary. Established in Law IV.

The Perceptory Boundary

The Perceptory Boundary is the structural element of a Perceptory Field that distinguishes its interior from its exterior. It is asymmetric, located at the closure-perimeter of the field, and dynamic rather than static, an active feedback loop that detects, compares, and adjusts to maintain the distinction between perceived and not-perceived. The Boundary is the structural location at which the Inseparation Constant operates. It is breachable: when two Perceptory Fields come into mutual registration, each field's Boundary is breached by the other, producing entanglement. The breachability of the Boundary is what permits Perceptory Fields to come into entangled relation, and is therefore the precondition of hierarchical organization across scales. Established in Law IV.

The Inseparation Element

The Inseparation Element is the structural object that holds an inseparable pair in their inseparable relation. Where the Inseparation Constant is the rule (every asymmetric Perceptory Field produces an inseparable pair at its boundary), the Inseparation Element is the structural object that the rule produces, the third element that constitutes the inseparation of the pair. At the matter scale, the Inseparation Element is the central element of the three-point closure; the framework predicts a specific instance, the A-particle, as the Inseparation Element of every particle/antiparticle pair. At other scales, the Inseparation Element appears as the magnetic field between magnetic poles, as the bond between mother and child, as the perceptory boundary between entangled and disentangled states for gravity, and ultimately as the Perceptory Force itself, the Inseparation Element of the two infinities. Established in Law VI; developed further in Laws VIII and X.

Hierarchical Entanglement

Hierarchical Entanglement is the architecture by which Perceptory Fields organize when they come into relation. The framework recognizes two relations, and only two. Entanglement is symmetric: two fields fused as peers into a single unit, neither containing the other. Containment is entanglement that has become asymmetric: one field held within another, so that a direction appears, an inside and an outside, a level. Hierarchy is nothing other than containment stacked upon containment, the directed form of entanglement rather than a principle imposed from outside. A network in which every field were merely entangled as an equal would be flat, perfectly symmetric, and therefore unstable, because the moment one field registers another field's registering of a third, asymmetry has entered, and with asymmetry, direction, and with direction, containment. Flatness collapses into hierarchy because asymmetry is unavoidable wherever perception compounds. Established in Law VI.

The Rule of Three

The Rule of Three is the geometric condition under which Hierarchical Entanglement produces stable Perceptory Fields. Three perceptory points entangled at a centre describe the minimum closed figure: a triangle whose interior exists by virtue of the three boundaries that define it. Three is the minimum number of entanglement points that produces a stable closed Perceptory Field; higher-order configurations are geometrically possible and are observed, where they occur, to decay toward three-point configurations; a formal stability ordering across configurations remains an open problem (Appendix B). In simple instances the three points configure as two inseparable opposites plus an Inseparation Element; in compound instances such as the strong force, the same Rule appears triadically rather than as a single linear pair, with the three-point geometry instantiated through three colour charges and their three anti-colours rather than through one opposite pair. The Rule of Three is not a feature of certain systems within reality but a structural condition for any stable existence whatsoever. Its clearest physical signature is the three-quark composition of hadrons, protons and neutrons, the constituents of all ordinary matter, are bound states of three quarks because three-point entanglement is the minimum stable Perceptory Field at the matter scale. Established in Law VI.

The Antipodal-Pair Rule

The Antipodal-Pair Rule is the structural rule that every fundamental force operating through finite-dimensional Perceptory Force modes exhibits antipodal-pair structure. The pair is not two independent forces but a single force with two complementary expressions, held in inseparable relation by an Inseparation Element at the centre between them. The Rule follows from the geometry of three-point closure: the two inseparable elements of a Perceptory Field must sit at antipodal positions, with the Inseparation Element balanced exactly between them, because antipodal placement is the only configuration in which two elements remain maximally distinct yet are held in stable relation by a mediating element that leans toward neither (Appendix B's balance condition). The Rule predicts that no fundamental force operating through a finite-dimensional Perceptory Force mode exists without an antipodal partner. Gravity is the structural exception, exhibiting antipodal-pair structure at a deeper level (entangling versus disentangling) rather than at the surface attraction-versus-repulsion level the other forces show. Established in Law VII; gravity's exception developed in Law VIII.

The Perceptory Network

A Perceptory Network is two or more Perceptory Fields in entangled relation through mutual breaching of their Perceptory Boundaries. When each field has registered the other within its own internal model, neither field can change without the change propagating to the other through the breach. Networks are the substrate of all observable interaction: at the matter scale, networks of particles produce the bound states of matter; at the cellular scale, networks of cells produce tissues and organs; at the cognitive scale, networks of cognitive Perceptory Fields produce shared meaning, social coordination, and consensus reality; at the cosmological scale, networks of massive Perceptory Fields produce galaxies, galaxy clusters, and the cosmic web. Gravity is defined in the framework as the act of perceptory entanglement between Perceptory Networks of the same structural kind. Term used throughout; gravity-relevant treatment established in Law VIII.

Perceptual Infinity

Perceptual Infinity is the totality of all perception observed as totality. It is symmetric, closed, and dimensionless. Its structure is the Perceptory Sphere of Appendix B: every perceptory distribution paired with its antipode under a pairing that carries no metric structure, infinite perceptory content without spatial extent. It is the same structural object as the Perceptory Field of Nothingness considered at the largest scale of the framework's architecture. Every perceptory point on the sphere of Perceptual Infinity is entangled with its antipode; this antipodal pairing at the level of the totality is the geometric origin of every opposite-pair phenomenon observed in physics, particle/antiparticle, north/south, charge polarity, the colour-and-anti-colour structure of the strong force, left-and-right-handed weak chirality. Established in Law X.

Propagation Infinity

Propagation Infinity is the totality of all asymmetric extension. It is open, ongoing, and never resolves into stable equilibrium. Its geometry is not the sphere but the ripple, perception extending across perception, breaking symmetry at every scale, propagating change without termination. Where Perceptual Infinity is the totality at rest, Propagation Infinity is the totality in motion. It is the one asymmetry of Law IX taken whole, held as mass, opened as space, expressed as force, and extended as gravity, gathered into a single totality of asymmetric extension. Time, in the framework, is what Propagation Infinity feels like to a Perceptory Network from inside: each network registers the asymmetries propagating through and around it as change, and the perception of that change is what we call the experience of time. Time is therefore substrate-relative, the time of a butterfly, of a human, and of a star are each local perceptions of Propagation Infinity by Perceptory Networks operating at their own substrates. Established in Law X.

Architectural note

The two infinities, Perceptual and Propagation, stand in antipodal-pair relation, held together by the Perceptory Force itself, which functions as the Inseparation Element of this largest of the framework's antipodal pairs. The Perceptory Force, the Inseparation Element of the two infinities, and the act of perception that grounds existence are not three different primitives. They are the same structural primitive named under different aspects depending on the work each Law is doing. This identification is the framework's settled architectural thesis: the closing position toward which the Ten Laws have been building, and the structural unity in which Part II's derivation finds its completion.

Appendix B, Mathematical Notes

This appendix presents the framework's mathematical apparatus as it currently stands. The reader should hold a specific orientation toward what follows. The material below works in four clearly separated registers, and each statement is labelled by register. **Definitions** introduce mathematical objects rigorously. **Propositions and theorems** are statements proved from the definitions; each is accompanied by a proof or proof sketch, and nothing is labelled a proposition unless it is actually derivable. **Postulates** are structural commitments the framework asserts without proof; they are the mathematical form of claims the prose develops, and they are marked as assertions. **Correspondences and open problems** connect the apparatus to existing physics and identify the work that remains; these are programmatic, not established.

This layering replaces the earlier draft's practice of presenting asserted identities in the notation of derived results. The framework's core claims are structural rather than quantitative, and most of its substantive content remains in the postulate register. What the definitional layer now provides is a stable ground on which genuine derivation is possible, and a small number of results below are derived in the full sense. The reader interested in formal development is invited to take this apparatus as a starting point. The reader interested only in the framework's structural claims may treat the appendix as supplementary and read past it without loss.

A note on notation. The universe of perceptory points is written $\mathcal{P}$. The registration relation of a Perceptory Field is written R (earlier drafts wrote ρ; the symbol is changed to avoid collision with the entanglement densities δ and $\hat{\delta}$ defined below). The Perceptory Force is written Ω throughout this appendix; the symbol Φ is reserved for integrated information in the sense of Integrated Information Theory, which this appendix and Appendix E discuss, and using one symbol for both inviting confusion.

Where the prose of the main text names the Perceptory Force without a symbol, Ω is the symbol this appendix supplies.

Section B.0, Formalization of the framework's primitives

What follows is a first formalization layer for the framework's primitive concepts. Each primitive is given a definition clear enough to support derivable propositions about its structure. The mathematics works in a hybrid set-theoretic and graph-theoretic register. The choice is deliberate: set-theoretic structure is sufficient to capture the framework's relational primitives (Perceptory Fields, Boundaries, Inseparation Elements), and graph-theoretic structure captures entanglement and Networks naturally. Category-theoretic generalization is the next formalization layer beyond this one and is flagged where appropriate but not undertaken here.

The universe. Let $\mathcal{P}$ be a non-empty set whose elements are perceptory points. The framework does not commit, at this layer, to whether $\mathcal{P}$ is finite, countably infinite, or uncountable; whether it carries topological or measure structure; or whether it has cardinal properties beyond non-emptiness. These commitments are introduced only where specific definitions require them, and every statement below that would require added structure (a topology, a metric, a vector-space structure) says so explicitly rather than borrowing the structure tacitly.

B.0.1, The Perceptory Field

Definition (Perceptory Field).** A Perceptory Field is a triple $F = (S, R, c)$ where: $S \subseteq \mathcal{P}$ is a non-empty subset of perceptory points (the field's content); $R \subseteq S \times S$ is a binary relation on S (the field's registration relation) that is reflexive (for all $p \in S$, $(p, p) \in R$) and symmetric (for all $p, q \in S$, $(p, q) \in R$ implies $(q, p) \in R$); and $c: S \to S$ is a closure map satisfying $c(c(p)) = c(p)$ for all $p \in S$ (idempotency) and $(p, c(p)) \in R$ for all $p \in S$.

The closure map c formalizes the framework's central structural commitment from Law III: a Perceptory Field is the structural object produced when perception closes a self-perceiving loop. The condition $(p, c(p)) \in R$ states that every point is registered with its closure-image. The idempotency condition states that further closure does not change what has already closed: closure is stable.

Proposition B.0.1.1 (the image of closure is its fixed-point set). For any Perceptory Field $F = (S, R, c)$, the image $c(S)$ equals the fixed-point set $\mathrm{Fix}(c) = \{ p \in S : c(p) = p \}$, and this set is non-empty.

Proof. If $q \in c(S)$, then $q = c(p)$ for some p, and $c(q) = c(c(p)) = c(p) = q$ by idempotency, so $q \in \mathrm{Fix}(c)$. Conversely if $q \in \mathrm{Fix}(c)$ then $q = c(q) \in c(S)$. Non-emptiness follows because S is non-empty and c is defined on all of S. ■

The framework names Fix(c) the closure-perimeter of the Field. The reader should note that this is a terminological choice motivated by the prose's boundary imagery, not a geometric result: without topological structure on $\mathcal{P}$, Fix(c) is simply a distinguished subset, and any perimeter-like reading is interpretive. The proposition's content is modest but real: closure always lands on, and stabilizes at, a non-empty distinguished subset of the field.

Definition (the Perceptory** Field of Nothingness).** The Perceptory Field of Nothingness is $F\emptyset = (\mathcal{P}, \mathcal{P} \times \mathcal{P}, \mathrm{id})$, the Perceptory Field whose content is the entire universe, whose registration relation is universal (every point registered with every point), and whose closure map is the identity. Note that the condition $S = \mathcal{P}$ is part of the definition; an earlier draft omitted it, which made the claimed uniqueness of F∅ fail.

The prose claims that Nothingness is perfectly symmetric and perfectly stable. The formalization layer can now state and prove a precise version of that claim. Perfect symmetry is formalized as invariance under every permutation of the universe: no perceptory point is distinguishable from any other.

Theorem B.0.1.2 (uniqueness of the ground state). Let $|\mathcal{P}| \geq 2$. Then F∅ is the unique Perceptory Field F = ($\mathcal{P}$, R, c) on the whole universe such that (i) R and c are invariant under every permutation σ of $\mathcal{P}$ (that is, (p, q) ∈ R if and only if (σ(p), σ(q)) ∈ R, and c(σ(p)) = σ(c(p))), and (ii) every point is registered with at least one point other than itself.

Proof sketch. The orbits of the permutation group acting on $\mathcal{P} \times \mathcal{P}$ are the diagonal Δ and its complement. A permutation-invariant relation is a union of orbits, so the invariant reflexive symmetric relations are exactly Δ and $\mathcal{P} \times \mathcal{P}$. Condition (ii) rules out Δ, forcing R = $\mathcal{P} \times \mathcal{P}$. For the closure map: suppose c(p) = q with q ≠ p. If $|\mathcal{P}| \geq 3$, choose r outside {p, q} and let σ be the transposition exchanging q and r; then σ(p) = p, so equivariance gives c(p) = σ(c(p)) = σ(q) = r, contradicting c(p) = q. If $|\mathcal{P}| = 2$, the only non-identity equivariant candidate is the swap, which is not idempotent. Hence c = id. ∎

This theorem is the appendix's formal rendering of the prose claim that the ground state is singled out by symmetry alone: among all Fields spanning the whole universe, perfect indistinguishability of points forces universal registration and identity closure, which is exactly F∅.

B.0.2, The Boundary as distinction operator

Definition (Boundary). Let F = (S, R, c) be a Perceptory Field. The Boundary of F, denoted ∂F, is the indicator function ∂F: $\mathcal{P} \rightarrow \{0, 1\}$ with ∂F(p) = 1 if p ∈ S and ∂F(p) = 0 otherwise. The Boundary is the distinction operator separating the field's interior S from its exterior $\mathcal{P} \setminus S$.

The earlier draft spoke of the Boundary's "discontinuity." That language is withdrawn at this layer: discontinuity presupposes a topology on $\mathcal{P}$, which has not been assumed. What the definition supports is the set-theoretic statement that ∂F partitions $\mathcal{P}$ into interior and exterior; the prose's claim that the Boundary is the structural location where interior meets exterior acquires geometric content only once $\mathcal{P}$ carries topological structure, which is deferred to a later formalization layer.

Definition (asymmetric Boundary). A Perceptory Field F has an asymmetric Boundary if and only if $S \neq \mathcal{P}$, that is, if and only if the field's content is a proper subset of the universe. F∅ has $S = \mathcal{P}$ and therefore has no asymmetric Boundary.

Definition (configuration and Boundary breach). A configuration is a pair $\mathcal{C} = (\mathcal{F}, \beta)$ where $\mathcal{F}$ is a set of Perceptory Fields and β assigns to each unordered pair $\{F_1, F_2\} \subseteq \mathcal{F}$ a (possibly empty) symmetric cross-registration relation $\beta(F_1, F_2) \subseteq (S_1 \times S_2) \cup (S_2 \times S_1)$. The Boundaries of F_1 and F_2 are mutually breached in $\mathcal{C}$ if and only if $\beta(F_1, F_2) \neq \emptyset$.

This replaces the earlier draft's definition, which spoke of a pair "entering" an extended relation, dynamical language with no dynamics defined. Breach is now a property of a configuration: a static, well-defined fact about whether cross-registration is present. Genuine dynamics, how configurations change, belongs to the open dynamics problem flagged in B.0.9. The structural point the prose makes survives intact: breach is mutual registration, not Boundary destruction. After breach both S_1 and S_2 persist; what the configuration adds is the cross-field relation $\beta(F_1, F_2)$.

B.0.3, The Inseparation Constant

Principle (Inseparation Constant, I^c).** For every Perceptory Field F = (S, R, c) with asymmetric Boundary, the interior S and the exterior $\mathcal{P} \setminus S$ are jointly non-empty, and neither can be removed while the asymmetric Boundary persists: collapsing the exterior ($S = \mathcal{P}$) yields a field of ground-state type with no asymmetric Boundary, and collapsing the interior ($S = \emptyset$) yields no Perceptory Field at all, since content is non-empty by definition.

Remark on logical status. As formalized, I^c is a consequence of the definitions: given that S is non-empty by definition and that asymmetric Boundary means $S \neq \mathcal{P}$, the joint non-emptiness of interior and exterior follows immediately. The earlier draft presented this biconditional as if it were a substantive formal result; it is not, and the framework gains nothing by obscuring that. The honest statement is this: the formal layer builds inseparability into the definition of an asymmetric Boundary, and the substantive content of the Inseparation Constant lies in the prose's modal claim, that interior and exterior come into existence together and cannot exist apart. Capturing that modal claim formally requires machinery (a modal or dynamical layer over configurations) that this formalization does not yet include. What the formal layer does deliver is the following exhaustive trichotomy.

Proposition B.0.3.1 (Boundary trichotomy). For any non-empty $S \subseteq \mathcal{P}$ exactly one of the following holds: $S = \mathcal{P}$, in which case any field on S is of ground-state type and has no asymmetric Boundary; or $\emptyset \subsetneq S \subsetneq \mathcal{P}$, in which case any field on S has an asymmetric Boundary and both interior and exterior are non-empty. The case $S = \emptyset$ supports no Perceptory Field. Consequently every Perceptory Field that is not of ground-state type has both a non-empty interior and a non-empty exterior. ■

The proposition formalizes the prose claim that the Inseparation Constant rules out genuine isolation for every Perceptory Field that is not the ground state: there is no Field whose interior exists without a non-empty exterior standing against it, except the ground state, which has no exterior because it has no asymmetric Boundary at all.

B.0.4, The Inseparation Element

The earlier draft defined the Inseparation Element with a closure condition that presupposed the Element and then "proved" that configurations lacking the Element fail the condition, a circularity. The definition below removes it by making mediation constitutive, and by distinguishing two graphs that the earlier draft conflated: the dependency graph, whose edges are the constitutive registrations, and the closure graph, which additionally represents the pair-relations those registrations constitute.

Definition (mediated pair and Inseparation** Element).**
A mediated pair is a triple (A, B, E) of pairwise disjoint non-empty subsets of $\mathcal{P}$ together with registration relations R_A ⊆ A × E and R_B ⊆ B × E such that every a ∈ A is registered with at least one e ∈ E, every b ∈ B is registered with at least one e ∈ E, and the balance condition holds: for every e ∈ E, the number of elements of A registered with e equals the number of elements of B registered with e (when these are finite; the measure-theoretic generalization is deferred). E is the Inseparation Element of the pair, written E_AB. The cross-pair relation is constituted by the Element: (a, b) ∈ R_AB if and only if there exists e ∈ E with (a, e) ∈ R_A and (b, e) ∈ R_B.

The balance condition replaces the earlier draft's centrality condition, whose qualifier "modulo the symmetry of the pair" was undefined and emptied the condition of content. Balance is exact and checkable: the Element does not lean toward either side of its pair.

Definition (dependency graph, closure graph). The dependency graph of a mediated pair has vertex set A ∪ E ∪ B and edge set R_A ∪ R_B. The closure graph additionally includes an edge (a, b) for every constituted pair (a, b) ∈ R_AB.

Theorem B.0.4.1 (the Element is the channel of the pair). In the dependency graph of a mediated pair, every path from a vertex in A to a vertex in B passes through E. Deleting E (and its incident edges) disconnects A from B entirely, and the constituted relation R_AB becomes empty.

Proof. By construction the dependency graph has no edge between A and B: all edges run between A and E or between B and E. Any A-to-B path therefore alternates through E. Deleting E removes all such paths, and R_AB is defined by existential quantification over E, so it is empty when E is. ■

The theorem is immediate from the definitions, and that immediacy is the point: the framework's claim is precisely that the relation between the members of an inseparable pair is constituted by the third element, so that dissolving the Element dissolves the relation. The earlier draft asserted this as Proposition B.0.4.1 in a form that was circular; the present form makes the constitution explicit in the definition and derives the dissolution claim from it.

Definition (stability). A finite registration graph is stable if and only if it is connected, contains at least one edge, and contains no bridge (an edge whose deletion disconnects the graph); equivalently, every edge lies on a cycle.

This supplies what the earlier draft lacked: an independent definition of stability against which minimality claims can be theorems rather than restatements. The motivation is the framework's own: a configuration held together by a registration whose loss would sever it is not stably closed, because closure means that every relation is secured by a loop.

Proposition B.0.4.2 (two-point configurations are unstable). A configuration consisting of two vertices and the single registration between them is not stable: its one edge is a bridge. ■

Theorem B.0.4.3 (the Rule of Three). Every stable registration graph has at least three vertices and at least three edges. The unique stable graph attaining both minima is the triangle K_3 (three vertices pairwise registered).

Proof sketch. A stable graph has an edge, and that edge lies on a cycle. Registration between distinct objects is irreflexive and symmetric, so the graph is simple and undirected, and every cycle in a simple graph has length at least 3; hence at least three vertices and three edges. A connected bridgeless graph on exactly three vertices with exactly three edges is the triangle, and the triangle is stable since each edge lies on the 3-cycle. ■

This is the formal content of Law VI's Rule of Three at this layer: among all configurations of mutually registered objects, the three-point closed cycle is the minimum stable structure, and the minimum is unique. Note what the theorem does and does not say. It derives the number three from the definition of stability and the simple-graph character of registration; it does not derive the physical claim that hadrons or other observed triads instantiate this structure, which remains a correspondence in the sense of this appendix's registers. Higher-order stable configurations exist (any cycle, any bridgeless graph); the theorem characterizes the minimum, and the earlier draft's unsupported claim that higher-order configurations are "energetically less stable" is withdrawn and reposed as an open problem (a stability ordering on configurations requires an energy functional the framework has not yet defined).

B.0.5, The Perceptory Network

Definition (Perceptory Network).** Given a configuration $\mathcal{C} = (\mathcal{F}, \beta)$, the Perceptory Network of $\mathcal{C}$ is the graph $\mathcal{N} = (\mathcal{F}, \mathcal{E})$ whose vertices are the Fields of $\mathcal{C}$ and whose edge set is $\mathcal{E} = \{ \{F_i, F_j\} : \beta(F_i, F_j) \neq \emptyset \}$, the pairs of Fields whose Boundaries are mutually breached. $\mathcal{E}$ is symmetric and irreflexive by construction.

Definition (entanglement density). The entanglement density of a Field F in a Network $\mathcal{N}$ is $\delta(F) = |N(F)|$, the number of Fields with which F is mutually breached (its degree in $\mathcal{N}$). This unweighted count is the defined base case. A weighted variant $\delta\_w(F) = \Sigma\, w(F, F_j)$ over the neighbourhood requires a specified weight function w capturing registration strength; no such function is defined at this layer, and results below use the unweighted case. The boundary-normalised entanglement density is $\hat{\delta}(F) = \delta(F) / |Fix(c\_F)|$ when Fix(c_F) is finite, the density per unit of closure-perimeter; its infinite-perimeter generalization is deferred.

Definition (subordination). In a Network $\mathcal{N}$, Field F_j is subordinate to Field F_i, written $F_j \preccurlyeq F_i$, if and only if $N(F_j) \setminus \{F_i\} \subseteq N(F_i) \setminus \{F_j\}$: every Field registered with F_j (other than F_i itself) is also registered with F_i. Subordination is the formal rendering of the prose claim that a hierarchically contained Field relates to the Network only through relations its container also holds.

Proposition B.0.5.1 (subordination is a preorder and forces the density ordering). The relation $\preccurlyeq$ is reflexive and transitive, and $F_j \preccurlyeq F_i$ implies $\delta(F_j) \leq \delta(F_i)$.

Proof sketch. Reflexivity is immediate. For transitivity, suppose $F3 \preccurlyeq F2$ and $F2 \preccurlyeq F1$, and let $x \in N(F3)$ with $x \neq F1$. If $x \neq F2$, then $x \in N(F2)$ by the first subordination and $x \in N(F1)$ by the second. If $x = F2$, then $F2 \in N(F3)$; $F2 \preccurlyeq F1$ requires $N(F2) \setminus \{F1\} \subseteq N(F1)$, and $F3 \in N(F2)$ with $F3 \neq F1$ gives $F3 \in N(F1)$, hence $F1 \in N(F3) \setminus \{F2\} \subseteq N(F2)$, hence $F2 \in N(F1)$ as required. Irreflexivity of $\mathcal{E}$ gives $x \neq F3$ throughout. For the density claim: if $F_i \in N(F_j)$ both neighbourhoods lose exactly one element under the set differences, and otherwise neither loses anything on the left, so the inclusion yields $|N(F_j)| \leq |N(F_i)|$ in both cases. ■

Remark. The implication runs from hierarchy to density, not the reverse. The earlier draft asserted the converse, that a raw density inequality $\delta(F_i) > \delta(F_j)$ by itself places F_j inside F_i's hierarchy; that claim is false in general (two Fields in disjoint regions of a Network can have any pair of densities whatever) and is withdrawn. What survives is the correct and provable direction: hierarchical subordination is always reflected in the density ordering, so entanglement density is a witness of hierarchy where hierarchy exists, not a generator of it.

B.0.6, The gravity analogue

Definition (substrate, entanglement magnitude). A substrate σ is an equivalence class of Perceptory Networks distinguished by the kind of entanglement they instantiate (matter-scale, cognitive, biological, cultural, and so on); $\mathcal{N}\_\sigma$ denotes the class of Networks at substrate σ. The entanglement magnitude of a Network is $M(\mathcal{N}) = \Sigma\ \delta(F)$ summed over its constituent Fields.

Definition (gravity analogue). The gravity analogue at substrate σ is a function $G\_\sigma : \mathcal{N}\_\sigma \times \mathcal{N}\_\sigma \rightarrow \mathbb{R}{\geq}0$ satisfying: (1) substrate restriction, $G\_\sigma(\mathcal{N}1, \mathcal{N}2) > 0$ only if both Networks are at substrate σ; (2) magnitude-monotone attraction, G_σ is monotonically non-decreasing in $M(\mathcal{N}1)$ and $M(\mathcal{N}2)$; and (3) antipodal-pair structure, $G\_\sigma = G\_\sigma^+ - G\_\sigma^-$ where G_σ^+ is an entangling component determined by the formation of cross-Network breaches and G_σ^- is a disentangling component determined by their dissolution, with the perceptory boundary between entangled and disentangled states (Law VIII) functioning as the Inseparation Element of the pair.

Remark on condition (3). A decomposition of a non-negative function into a difference of two non-negative functions is trivially available and carries no content by itself. Condition (3) is therefore stated with a non-triviality requirement: G_σ^+ and G_σ^- must be independently specified by the entangling and disentangling processes respectively (breach formation and breach dissolution in the configuration dynamics), so that the decomposition is fixed by the structure rather than chosen freely. Absent a dynamics, this requirement is a constraint on future formalization rather than a present construction; it is recorded so that the decomposition cannot be satisfied vacuously.

Remark (accumulative, not antipodal). Condition (2) is the formal signature of what Law VIII calls accumulative attraction: G_σ is monotone in magnitude and never reverses sign, so the gravity analogue has no opposite that cancels it. This is precisely what distinguishes it from the antipodal forces of Law VII, whose coupling pairs across a sign; those sign-paired attractions are not gravity analogues and lie outside the scope of G_σ. The antipodal structure invoked in condition (3) is gravity's own deeper pair, entangling against disentangling, not the sign-paired attraction of Law VII.

Correspondence (matter substrate). For σ = matter, the framework asserts that G_matter is the established gravitational interaction. In the Newtonian regime the asserted correspondence is $G\_matter(\mathcal{N}1, \mathcal{N}2) = \kappa \cdot M(\mathcal{N}1) \cdot M(\mathcal{N}2) / d^2(\mathcal{N}1, \mathcal{N}2)$, where d is spatial separation and κ is a substrate conversion constant carrying the units required to map entanglement magnitude to mass; the identification of matter-scale entanglement magnitude with mass-energy is an asserted correspondence of the framework, not a derivation, and the existence of a well-defined κ is part of what the correspondence asserts. The earlier draft wrote this law in terms of densities on both sides, which mismatches the Newtonian form (which couples masses, not densities) and is corrected here. In strong-field regimes the correspondence is asserted qualitatively with the curvature description of general relativity; no formal map is offered at this layer.

Open problem (coupling functions and non-matter substrates). The functional form of G_σ for σ ≠ matter is unspecified. The framework predicts that G_σ exists with properties (1) through (3); whether the attraction follows an inverse-square law, another power law, an exponential decay, or a substrate-specific form is a question that deriving G_σ from the structural primitives must answer. The framework does not commit to inverse-square form at non-matter substrates, because the structural reasons for inverse-square behaviour at the matter substrate (three-dimensional spatial geometry, conservative-field propagation) do not transfer to substrates where distance may not be spatial. A notion of distance d for non-matter Networks must itself be defined, plausibly as a graph metric on the Network; this too is open.

B.0.7, Relations among the primitives

The primitives defined above are not independent. The following relations connect them; each is labelled with its register.

Relation 1 (Boundary as locus of the Inseparation Constant; proposition). For any Perceptory Field with asymmetric Boundary, the trichotomy of B.0.3.1 holds at that Boundary: interior and exterior are jointly non-empty exactly where the Boundary distinguishes them.

Relation 2 (the Element holds its pair; theorem). For every mediated pair, the Inseparation Element is the unique channel of the cross-pair relation, by Theorem B.0.4.1.

Relation 3 (Networks from breach; definition). Network edges are exactly mutual Boundary breaches within a configuration; Networks are constructed from breach relations among Fields.

Relation 4 (gravity from Network structure; postulate plus definition). G_σ is determined by the entanglement magnitudes of its arguments, which are graph-theoretic properties of Networks; gravity at any substrate is therefore a derived consequence of Network structure rather than a primitive, conditional on the correspondence of B.0.6.

Relation 5 (the Perceptory Force as universal Inseparation Element; postulate). The Perceptory Force Ω is the Inseparation Element of the largest mediated pair the framework recognizes, perceptual infinity I_p and propagation infinity I_π: $\Omega = E_{I_p I_\pi}$. The identification types Ω as a structural object in the sense of B.0.4, consistent with its architectural role in Appendix A. This is the framework's settled architectural thesis, asserted, not derived.

B.0.8, Case study: the A-particle and the structure of unmeasurable predictions

The A-particle prediction (Law VI; Appendix C, Prediction 1) provides a case study in how the framework handles predictions whose target is constitutively unmeasurable as a free-state entity but measurable through its effects on observable quantities.

The structural setting. The A-particle is the Inseparation Element of every particle-antiparticle pair (Law VI). As Inseparation Element it occupies the mediating position of a three-point closed structure whose other two nodes are the particle and the antiparticle. Any registration event that would measure the A-particle as a separate entity introduces a fourth perceptory entanglement into the three-point structure, reconfiguring it and collapsing the very object that registration would attempt to detect. The A-particle is therefore constitutively unmeasurable as a free-state particle in the framework's reading;

this is a structural consequence of Law VI's architecture, not a contingent fact about measurement technology.

The measurable target, restated. The earlier draft wrote the prediction as an energy decomposition E_total = E_QED + E_A with E_A > 0, which as stated conflicts with energy conservation: the total energy of a pair-annihilation event is fixed by the rest masses and kinetic energies of the annihilating pair, and quantum electrodynamics does not leave an unallocated remainder in the total. The corrected statement locates the prediction where it belongs: not in the size of the total energy budget but in its distribution. For an annihilation observable O (a spectral shape, an angular correlation, a branching ratio, a timing distribution), define the anomaly A(O) = O_observed − O_QED, the deviation of the measured observable from the complete QED prediction at fixed total energy. The framework's prediction is that the dissolution of the A-particle's mediating role during annihilation leaves a signature in at least one class of observables: A(O) ≠ 0 for some O, with the deviation systematically associated with the three-point-to-two-point reconfiguration. Falsification consists in showing that all annihilation observables agree with QED at the highest attainable precision.

The precision constraint,** stated honestly.** QED's predictions for the relevant systems are confirmed to extraordinary precision; positronium decay rates and energy levels, and QED precision tests generally, agree with theory at the parts-per-million to parts-per-billion level depending on the observable. Any A-particle signature must therefore live below current experimental bounds or in observables not yet tested with precision. This sharply constrains the prediction and is acknowledged rather than elided: until the framework supplies a predicted magnitude for A(O), the prediction is qualitative, and a null result at any given precision lowers the allowed magnitude without strictly falsifying the qualitative claim. Converting the qualitative prediction into a falsifiable quantitative one is precisely the theoretical work the framework calls for, and the framework accepts that the prediction's scientific weight remains limited until that work is done.

Required formal properties. To convert the qualitative prediction into a quantitative experimental target, theoretical work must characterize the A-particle's contribution to annihilation observables through seven effective properties: effective mass m_A, effective spin s_A, effective charge q_A, effective lifetime τ_A, effective coupling g_A, effective decay channel D_A, and conservation compatibility C_A, the requirement that all Standard Model conservation laws (energy, momentum, charge, baryon number, lepton number) remain satisfied across the reconfiguration, with the A-particle's contributions integrated into the total ledger rather than added on top of it. These properties cannot be measured directly, since the A-particle is unmeasurable as a free state; they are theoretical parameters through which the anomalies A(O) would be computed.

Status and generalization. The framework currently supplies the structural account and the qualitative prediction; it does not supply the decompositions through which the seven properties could be extracted from observables. The case-study structure generalizes: Inseparation Elements at other substrates (the bond, the Perceptory Boundary as Element-location, the Perceptory Force as Element of the two infinities) are similarly unmeasurable as free states by the same architecture, and their measurable targets are likewise distributional signatures rather than additions to conserved totals.

B.0.9, What this formalization layer has established

The layer gives the framework's primitives definitions clear enough to support derivation, and it now contains results that are proved rather than asserted: the image of closure is its non-empty fixed-point set (B.0.1.1); the ground state is the unique fully symmetric Field on the universe (B.0.1.2); the Boundary trichotomy (B.0.3.1); the Element-as-channel theorem (B.0.4.1); the instability of two-point configurations (B.0.4.2); the Rule of

Three as a minimality theorem with a unique minimum (B.0.4.3); and the subordination preorder with its density consequence (B.0.5.1). Alongside these, the layer states its postulates and correspondences as such: the identification of matter-scale entanglement magnitude with mass-energy, the existence and properties of G_σ at non-matter substrates, and the identification $\Omega = E\_\{I\_p\ I\_\pi\}$.

What remains is the next formalization layer: dynamics (how configurations evolve, which would give content to breach formation and dissolution and to the G_σ^+/G_σ^- decomposition), couplings (the form of G_σ at each substrate and the conversion constant κ), an energy functional (which would turn the stability ordering of configurations into a derivable claim), and the connection to existing physics through the action principle below. The sections that follow are kept deliberately closer to the prose register and are marked according to the same four registers used throughout.

The Perceptory Sphere

The framework's central geometric object is the Perceptory Sphere, the structural representation of perceptual infinity that Law X developed: the totality of all possible perceptory distributions perceiving Nothingness simultaneously, with each distribution paired against its opposite.

Definition (Perceptory** Sphere).** The Perceptory Sphere is a pair $\mathbb{S} = (\Pi, *)$ where Π is the set of all possible perceptory distributions and $*: \Pi \to \Pi$ is the antipodal map, a fixed-point-free involution: $p** = p$ for all $p \in \Pi$, and $p* \neq p$ for all $p \in \Pi$.

Proposition (antipodal pairing partitions the Sphere). The relation $p \sim q$ defined by $q = p$ or $q = p*$ is an equivalence relation whose classes are exactly the unordered pairs $\{p, p*\}$. Consequently Π is partitioned into antipodal pairs, no distribution is its own opposite, and Π is either infinite or of even finite cardinality.

Proof. Reflexivity is immediate; symmetry follows from the involution property; transitivity follows by cases. Each class has exactly two elements because $*$ is fixed-point-free. ∎

Remark on the sphere language and on the earlier formulation. The geometric sphere is the familiar object that carries a fixed-point-free antipodal involution, and the framework borrows the name for exactly that structural feature: every point paired with an opposite, the totality closed under the pairing. What the framework's Sphere does not carry is any metric, topology, or embedding in extended space, because on the framework's account space does not yet exist at the level of perceptual infinity. The earlier draft expressed this by writing $V = 0$ and $r = 0$ and by substituting "infinite perceptory points" into the Euclidean sphere equation. Those formulations are withdrawn: a radius-zero Euclidean sphere is a single point, which is the opposite of what the framework intends, and the substitution of a cardinality into a coordinate slot is not a meaningful equation. The intended content is stated directly instead: $\mathbb{S}$ carries antipodal structure without metric structure. The claim that perceptual infinity has no spatial extent is formalized as the absence of any metric on Π at this layer, not as a zero value of a metric quantity.

The three identities

The framework identifies three structural constants operating at every scale of the perceptory hierarchy. The earlier draft presented them as bare symbolic equations; each is now a defined quantity, so that the assertions have content and, in one case, the asserted value becomes a theorem.

Unity of perceptual infinity (I(id) = 1). Let the total Network be the Perceptory Network of the configuration comprising all Perceptory Fields and all breaches. Define I(id) as the number of connected components of the total Network. The framework postulates I(id) = 1: perceptual infinity is structurally singular, all Fields belong to one connected whole, none stands in absolute isolation. This is a postulate; it is the graph-theoretic form of the prose claim that perceptual infinity is unified, and Proposition B.0.3.1 supports but does not entail it.

Completeness of coexistence (C(id) = 1). Define C(id) = 1 if every perceptory point of $\mathcal{P}$ belongs to the content of at least one Perceptory Field, and C(id) = 0 otherwise. The framework postulates C(id) = 1: nothing is excluded from coexistence at the level of perceptual infinity. This too is a postulate, the set-theoretic form of the prose claim that coexistence is complete.

Rule of Three (H(id) = 3). Define H(id) as the number of vertices of the minimum stable registration graph in the sense of B.0.4. Then H(id) = 3 by Theorem B.0.4.3, and the minimum is uniquely attained by the triangle. Of the three identities this one is now a theorem of the formal layer rather than a postulate: the Rule of Three is derived from the definition of stability and the simple-graph character of registration.

The relationship among the identities is structural. I(id) = 1 and C(id) = 1 together state that perceptual infinity is unified and complete in its coexistence; H(id) = 3 states that the condition of stable closure within it is three-point structure. The first two are commitments of the framework; the third is what the formal layer can already deliver on its own.

Antipodal pairing

The structural geometry of Law VII's Antipodal-Pair Rule is captured by combining the Sphere's involution with the Inseparation Element of B.0.4. For a distribution P ∈ Π, its antipode is P∗, and the pair {P, P∗} is held by its Inseparation Element E_{PP∗} in the sense of B.0.4: a mediating object through which the pair's relation is constituted, balanced exactly between the two sides, whose removal dissolves the relation (Theorem B.0.4.1). This typing is deliberate and corrects an inconsistency in the earlier draft, which defined the Element as a set of perceptory points in B.0.4 but then, in this section, as the coordinate midpoint E = (P + P∗)/2 = 0 of a vector representation. The two definitions are incompatible, and the second collapses every pair's

Element to one and the same point, the origin, erasing the distinctness of Elements that the framework everywhere else requires.

The vector picture can be retained for what it is: an illustration. In any representation of Π in a vector space in which the antipodal map is represented by negation, each pair {P, −P} has midpoint at the origin. The representation therefore depicts the Element's exact balance between its pair (the content of the balance condition of B.0.4) at the cost of mapping all Elements to a single representative point; it is a picture of one structural property, not a definition of the Element. Similarly, the earlier draft's centroid formula $E = (P_1 + P_2 + P_3)/3$ for three-point closure depicted the Element's centrality, but the centroid of a triangle is not in general equidistant from its vertices (that is the circumcenter), so the formula did not even illustrate the equidistance it was meant to capture. In the present apparatus centrality is carried by the balance condition of the mediated-pair definition, which is representation-free, and three-point closure is carried by Theorem B.0.4.3.

The action principle (speculative formalization)

The framework's most speculative formalization gesture is the proposal that the Perceptory Force introduces an additional term in the action principle of a physical system. In standard physics the action S of a system is the time integral of its Lagrangian, with dynamics determined by the variation of S. The framework proposes

$$S\_total = S\_gauge + S\_PF[\psi(t), \partial\_t\, \psi(t)]$$

where S_gauge is the action of the system's gauge interactions (electromagnetism, strong force, weak force, and the gravitational-curvature contribution) as standard physics describes them, and S_PF is the proposed Perceptory-Force contribution. The argument ψ(t) is the system's internal state-vector, the carrier of identity-entanglement, in the framework's vocabulary the structural representation of the Perceptory Field's self-perception.

(The symbol ψ here is the state-vector argument of the action and is unrelated to the Perceptory Force symbol Ω; and Φ, as noted at the head of this appendix, is reserved for integrated information.)

The proposed structure of S_PF has three features, all programmatic. First, a decoupling constraint: when the Perceptory-Field loop is broken, no coherent ψ(t) exists and S_PF must vanish identically, so that gauge-only physics re-emerges untouched; formally, S_PF → 0 in the limit of dissolved Field coherence. This constraint is what protects experimentally validated physics in regimes where Field coherence is not the relevant feature, and any candidate S_PF failing it is excluded. Second, in the coherent regime S_PF modulates the gauge sector through a coupling the framework has not specified; the general structure is the Sense, Compare, Act feedback of Law IV, proposed here as a modification of the system's action. Third, the modification is structurally local: it operates at the Boundary of the Perceptory Field, not throughout its interior, consistent with Law IV's localization of the Inseparation Constant and its identification of the Boundary as where the feedback loop closes. To these the present revision adds an explicit conservation constraint: S_PF must either preserve the conservation laws of the Standard Model exactly or predict specific, quantified deviations; a modification that silently violates conservation is not a candidate.

Several developments are required before the formulation can be tested. The coupling, the precise mathematical form by which S_PF modulates the gauge sector, must be specified, either by derivation from the framework's structural principles or as a phenomenological proposal. The state-vector ψ(t) must be characterized: whether it lives in a Hilbert space analogous to quantum-mechanical state vectors or in a different category appropriate to the framework's commitments is open. And the coherence criterion, the quantitative condition distinguishing a coherent loop from a broken one, must be supplied, whether developed natively or adapted from existing measures (integrated information Φ in the sense of IIT, free energy in the sense of the free-energy principle). This is a programme of theoretical-physics work, not a completed result; the formulation is offered as a target that future work may develop, refine, or replace.

What this appendix has established

The apparatus now separates its registers explicitly. Proved at this layer: closure stabilizes on a non-empty fixed-point set; the ground state F∅ is the unique fully symmetric Field on the universe; every non-ground-state Field has non-empty interior and exterior; the Inseparation Element is the constitutive channel of its pair, whose removal dissolves the pair's relation; two-point configurations are unstable; the minimum stable configuration is the three-point cycle, uniquely; hierarchical subordination is a preorder that forces the entanglement-density ordering; and the Perceptory Sphere's antipodal map partitions perceptual infinity into pairs with no self-paired distribution. Postulated: the unity and completeness identities I(id) = 1 and C(id) = 1; the existence and properties of the gravity analogue at non-matter substrates; the identification of the Perceptory Force as the Inseparation Element of the two infinities. Asserted as correspondence: the identification of matter-substrate entanglement magnitude with mass-energy and of G_matter with established gravitation. Open: dynamics, couplings, the energy functional, the distance notion for non-matter substrates, the quantitative content of the A-particle prediction, and every component of the action-principle programme.

The apparatus serves the same three purposes as before, now on firmer ground. It makes the framework's structural commitments specific. It provides a target for formal development, with the postulates and open problems identifying exactly where rigorous work must enter. And it offers readers trained in mathematics and physics an interface into the framework's claims in which what is proved, what is asserted, and what is open are never allowed to blur. The framework does not require this apparatus to be final. It requires only that future work clarify, refine, or replace each element, with the structural claims surviving or being revised according to what that work reveals. The mathematics is offered for engagement, not for closure.

Appendix C, Predictions and Falsifiability

The framework of the Laws of Consciousness is not a metaphysical proposal sealed off from empirical disconfirmation. It is a structural theory whose claims commit to specific testable consequences. This appendix consolidates the framework's predictions across the Ten Laws and identifies, for each prediction, the condition under which the framework would be falsified at that point. Some predictions are testable with current technology; others require advances in instrumentation or in cosmological observation; one or two are testable only in principle, given the current state of the field. None is exempt from disconfirmation. A framework whose predictions cannot fail is not a framework, it is decoration.

The predictions below are listed in order of the Laws from which they derive.

1. The A-particle as Inseparation Element of particle-antiparticle pairs

Source: Law VI. Status: Sharp, falsifiable through the collapse-signature observable, testable in principle with sufficiently precise measurement of pair-annihilation energetics.

The framework predicts that for every particle-antiparticle pair, there exists a third structural element, the A-particle, that constitutes the pair's inseparation, with the A-particle functioning as the Inseparation Element of the three-point Perceptory Field that the particle, the antiparticle, and the A-particle together form. Pair-annihilation is therefore not the simple mutual cancellation of particle and antiparticle; it is the collapse of the three-point Perceptory Field whose centre is the A-particle, and the distribution of pair-annihilation's fixed energy budget across observables should reflect the A-particle's mediating role in that Field. The framework predicts that this role is detectable as distributional anomalies in pair-annihilation events, deviations in spectral shapes, angular correlations, branching ratios, or timing at fixed total energy, relative to the complete predictions of standard quantum electrodynamics. The A-particle itself is constitutively unmeasurable as a free-state entity, measurement would reconfigure the three-point structure that the A-particle holds together, but the collapse signature is measurable. The framework's prediction is therefore a prediction about anomalies in pair-annihilation observables, not about a directly detectable third particle.

Falsification condition. If high-precision pair-annihilation experiments, particularly electron-positron annihilation under controlled conditions, confirm that all pair-annihilation observables agree with the complete predictions of standard quantum electrodynamics at the highest precision attainable, the prediction is progressively constrained; the framework acknowledges that, until it supplies a predicted magnitude, null results bound the anomaly without strictly falsifying the qualitative claim, and that the prediction's scientific weight remains limited until that quantitative work is done. The framework's reading of pair-annihilation as three-point Perceptory Field collapse, with the A-particle as Inseparation Element, would then need to be revised at this point. If anomalies are detected at sufficient statistical significance, the framework's A-particle interpretation becomes a candidate explanation requiring formal modelling and independent confirmation. Detection of an anomaly does not, on its own, establish the A-particle as a confirmed structural object. The standard sequence applies: anomaly identification, replication under controlled conditions, exclusion of known alternative causes, formal-theoretical modelling that links the anomaly's specific signature to the A-particle's predicted role as Inseparation Element, and independent confirmation by groups not affiliated with the original detection. Only after this full sequence would the A-particle interpretation move from candidate explanation to established theoretical object. The next step in the framework's research programme, given anomaly detection, is the development of theoretical decompositions through which the A-particle's structural properties (mass, spin, charge, coupling, lifetime, decay behaviour) could be extracted from the collapse-signature observable as part of that formal-modelling work.

2. Weak-force parity restoration at high energy

Source: Law VII. Status: Sharp, falsifiable, testable through high-energy collider experiments and early-universe cosmological signatures.

The framework predicts that parity violation in the weak interaction, the asymmetric coupling of the weak force to left-handed particles, with right-handed coupling suppressed, is amplitude asymmetry under current low-energy conditions, not antipodal absence. Both antipodes (left-handed and right-handed weak interactions) exist; the right-handed antipode is suppressed at currently accessible energy scales. At sufficiently high energy regimes, energies approaching unification scales or those obtaining in the very early universe, parity should be restored, with left- and right-handed weak interactions exhibiting equal amplitudes.

Falsification condition. If parity remains permanently broken at all accessible energy scales, with no convergence toward restoration as energy increases, and if early-universe cosmological signatures consistent with restored parity also fail to appear, the prediction is falsified. The framework's reading of parity violation as amplitude asymmetry (rather than antipodal absence) is wrong, and this section must be revised.

3. No genuine isolated magnetic monopole

Source: Law IV. Status: Sharp, falsifiable through ongoing monopole searches.

The framework predicts that no truly isolated magnetic monopole will ever be observed. Magnetic poles are an inseparable pair generated by the Inseparation Constant operating at the electromagnetic substrate; the pair cannot be structurally severed without dissolving the field that produces it. What appears as a local monopole, in any frame where the framework's prediction is to be tested, must have its inseparable opposite somewhere within the larger Perceptory Field, perhaps at great distance, perhaps at cosmological separation, but present in the field as a structural necessity.

Falsification condition. Confirmed, replicated direct observation of a free fundamental magnetic monopole, detected as an isolated particle with no paired opposite and no compensating field structure within the experimentally characterized system, would falsify the prediction and require revision of Law IV. The 1982 contested detection does not count as confirmed; nor would brief detection of a quasi-particle excitation analog. The prediction concerns fundamental monopoles in the strict sense.

4. Cosmic web structural regularities beyond random gravitational collapse

Source: Law IX. Status: Sharp, falsifiable through detailed cosmological observation against ΛCDM predictions.

The framework predicts that the filamentary geometry of the cosmic web is the visible signature of asymmetric perceptory entanglement distribution at the cosmological scale. Three specific testable claims follow. First: voids will exhibit structural regularities consistent with perceptory symmetry conditions rather than statistical randomness. Second: filaments will exhibit hierarchical self-similarity consistent with the Rule of Three operating at cosmological scale, including specific angles of intersection and characteristic length-ratios across scales. Third: the geometric relationships among filaments and voids will reveal the underlying entanglement-asymmetry distribution rather than purely gravitational evolution from random initial conditions.

Falsification condition. If detailed cosmic-web observation reveals only what standard ΛCDM cosmology predicts, purely statistical structure consistent with gravitational collapse from random primordial fluctuations, with no structural regularities beyond what gravity alone produces, the prediction is falsified. The framework's reading of the cosmic web is wrong, and the cosmological-scale claims of Law IX must be revised at this point.

5. Substrate-specific gravitational analogues

Source: Law VIII. Status: Sharp prediction; falsifiable in principle, though direct observation at non-matter substrates requires methodologies that do not yet exist.

The framework predicts that accumulative attraction, the opposite-less drawing-together that Law VIII distinguished from the antipodal attraction of Law VII, operates at every scale of hierarchical entanglement, with the same structural mechanism (perceptory entanglement between Perceptory Networks of the same structural kind) on the substrate appropriate to that scale. Matter-scale gravity is what mainstream physics describes; cognitive-scale accumulation (cognitive networks gravitating cognitive content), biological-scale accumulation (ecosystems accumulating species, immune systems accumulating pathogenic recognition), cultural-scale accumulation (intellectual traditions drawing practitioners, ideologies accumulating adherents, languages absorbing other languages), and other substrate-specific accumulative dynamics are predicted to be the same mechanism at different substrates, with attraction proportional to total entanglement at each substrate. The antipodal attractions at those substrates, which pair across a sign, belong to Law VII and are not gravity analogues.

Falsification condition. Discovery that accumulative attraction operates only at the matter scale, with no measurable analog at other substrates, would falsify this commitment. The framework's claim of substrate-dependence would be wrong, and Law VIII would reduce to a substrate-bound restatement of Verlinde's entropic-gravity programme rather than the universal substrate-extending claim it now makes.

6. Black holes as Inseparation Element collapse, with bipolar jets as the disentangling phase

Source: Law VIII. Status: Sharp prediction; testable through observation of black hole accretion and the bipolar jets it launches, with the quantum-scale counterpart testable only in principle.

The framework predicts that black holes are the visible signature of an Inseparation Element collapse at the matter scale, with the in-pulling of matter constituting the entangling phase and the bipolar jets the collapse launches constituting the visible disentangling phase, with Hawking radiation as the quantum-scale counterpart. The two phases are not separate phenomena governed by different physics; they are two phases of the same antipodal-pair structure of gravity (entangling/disentangling) under critical stress, the in-fall and the outflow held in inseparable relation by the perceptory boundary that is the Inseparation Element of gravity.

Falsification condition. If detailed observation of black hole accretion and jet-launching reveals that the energy budget of inflow and outflow corresponds purely to mass-energy conservation as currently formulated, with no signature consistent with Inseparation Element reconfiguration dynamics, the prediction is falsified. Specific markers to look for include: correlations between accretion (the entangling phase) and jet launching (the disentangling phase) that exceed what current magnetohydrodynamic models predict; anomalous structure in jet energetics suggesting reconfiguration rather than purely electromagnetic-mechanical driving; at the quantum-scale counterpart, anomalous spectral structure in Hawking radiation suggesting reconfiguration rather than thermal emission; and conditions under which black holes disperse rather than persist.

7. No genuine vacuum at any scale

Source: Law IX. Status: Committed; disconfirmable in stages as measurement sensitivity increases at every observable scale.

The framework predicts that what appears as empty space is always structured by entanglement asymmetries at the appropriate substrate. This claim is universal across scales: the empty space inside an atom is structured by atomic-substrate asymmetries; the space between cells is structured by biological-substrate asymmetries (extracellular matrix, signalling molecules); the space between people is structured by social-substrate asymmetries (communicative, cultural, informational); the cosmological vacuum is structured by cosmological-substrate asymmetries.

Falsification condition. Demonstration that established vacuum structure (zero-point fluctuations, vacuum energy, field condensates) is absent from some region at the scales accessible to measurement would falsify the prediction; since absence cannot be confirmed outright at finite sensitivity, the framework accepts that the prediction is disconfirmed in stages, with each empty layer of structure counting against it. The framework's claim that no Perceptory Field is empty would be wrong at that scale, and Law IX would need to be qualified accordingly.

8. Three spatial dimensions at every observable scale

Source: Law IX. Status: Sharp, falsifiable through direct observation if extra spatial dimensions exist.

The framework predicts that no directly observable extra spatial dimensions exist at any scale humans can measure, with three observable spatial dimensions derived from three independent symmetry breaks within the Perceptory Field at the scale of observation. The appearance of higher dimensionality in mainstream theoretical physics, string theory's ten or eleven dimensions, M-theory's eleven, various extensions, is read by the framework, where the dimensions are inferred from formal mathematics rather than observed, as one of three misreadings: time misread as spatial, formal-mathematical artifact, or non-spatial perceptory nesting confused with spatial geometry.

Falsification condition. If extra spatial dimensions are ever directly observed, measured as physical extension rather than inferred from formal mathematics, the framework is falsified at this point. The misreading classifications above apply only to formal-mathematical inference and may not be invoked against a confirmed direct observation. Direct observation of a fourth or fifth spatial axis through which physical entities can demonstrably move would invalidate the framework's commitment to three spatial dimensions and require substantial revision of Law IX.

9. Substrate-relativity of time

Source: Law X. Status: Sharp prediction extending relativity; testable through detailed comparative measurement of perceptual-loop rates across substrates.

The framework predicts that time is substrate-relative, that each Perceptory Network registers change at the rate of its own perceptual loops, with the time of a butterfly, a human, and a star being local perceptions of Propagation Infinity at different substrates, none privileged or universal. This goes beyond relativistic time dilation, which describes how time varies for one substrate (the matter scale) under different velocities and gravitational conditions. The framework predicts that time also varies between substrates in ways relativity does not address.

Falsification condition. Demonstration that the perceptual-loop rates of different substrates are reducible to a single underlying time-rate, that butterfly-time, human-time, and stellar-time are not structurally different timescales but the same time at different reference frames, would falsify the substrate-relativity claim. The framework's reading of time would reduce to standard relativistic time, and Law X's substrate-extending claim would be wrong.

10. The third element in the atomic binding structure

Source: Law VI. Status: Speculative prediction; testable in principle through high-precision electron-field measurement.

A subsidiary prediction follows from the Rule of Three operating at the atomic scale. The framework predicts that the nucleus-electron binding is mediated by a third structural element, the Inseparation Element of the nucleus-electron pair, satisfying the three-point closure that the Rule of Three requires for atomic stability; whether this element is identifiable with the electromagnetic binding field or constitutes a distinct structural object is left open. This prediction is more speculative than the others because the rule's atomic-scale instantiation has not yet been worked out in detail.

Falsification condition. Confirmed demonstration that atomic stability is fully accounted for by proton-neutron pairing alone, with no third structural constituent required and no anomalies in atomic-scale measurement that would be consistent with one, would falsify the prediction. The framework's claim that the Rule of Three operates at every scale of the hierarchy would need to be qualified to exclude the atomic scale, requiring partial revision of Law VI.

Architectural note on falsifiability

The framework's predictions vary in their immediacy of testability. Predictions 2, 3, and 4 are testable with current or near-future technology and observation; prediction 1 is testable in principle, with current experiments bounding the anomaly until the framework supplies a quantitative magnitude; predictions 5, 7, 9, and 10 are testable in principle but await methodological developments; prediction 6 is testable through current observation of black hole accretion and jet dynamics, with its quantum-scale counterpart testable only in principle; prediction 8 is falsifiable by direct observation should the relevant capability ever exist. None of the predictions is metaphysical in the sense of being immune to disconfirmation. Each can fail. The framework offers itself for failure at each of these points and accepts the consequences of any failure that occurs.

A reader who wishes to engage with the framework critically is invited to identify additional points at which it commits to specific consequences, to develop the falsification conditions more sharply than this appendix does, or to design experimental tests that would settle each prediction. The framework's seriousness as a theory is measured by its willingness to accept such engagement, and by the precision with which it makes itself available to be wrong.

Appendix D, Glossary

This glossary provides quick-reference definitions for the technical vocabulary used throughout the book. It is alphabetically ordered. Where a term has been formally defined in Appendix A, the glossary entry is brief and refers the reader to the full formal definition. Where a term appears only in passing within the book, the glossary entry is self-contained.

A-particle. The framework's predicted Inseparation Element of every particle-antiparticle pair, holding the pair in inseparable relation at the matter scale. Predicted to leave distributional signatures in pair-annihilation observables, deviations at fixed total energy from the complete predictions of quantum electrodynamics. See Law VI; Appendix C, Prediction 1.

Acceleration / deceleration. A paired field-formation dynamic by which Perceptory Fields increase or decrease the rate at which their feedback loops close, producing substrate-relative variations in their local perception of time. See Chapter 3.

Amplification / diffusion. A paired field-formation dynamic by which Perceptory Fields concentrate their intensity into fewer stronger entanglements (amplification) or spread it across more, weaker entanglements (diffusion). The two are structurally exclusive at any given moment. See Chapter 3.

Antipodal-Pair Rule. The structural rule that every fundamental force operating through finite-dimensional Perceptory Force modes exhibits antipodal-pair structure. See Appendix A; Law VII.

Asymmetric / symmetric processing. A paired field-formation dynamic by which Perceptory Fields break or preserve symmetry in their operation. Almost all observable existence is asymmetric processing; symmetric processing occurs only in the Perceptory Field of Nothingness or in conditions where symmetry has been deliberately preserved. See Chapter 3.

Bond, the. The Inseparation Element of the mother-child pair at the biological-relational scale; a real perceptory entity, not a metaphor. Severance produces a wound, not separation. See Law VI; Chapter 2.

Boundary breach. The condition under which two Perceptory Fields come into entangled relation: each field's Perceptory Boundary is penetrated by mutual perception, with each field having registered the other within its own internal model. The Boundary persists; mutual registration produces entanglement. See Law V.

Coexistence Constant (C(id) = 1). A structural postulate (Appendix B) asserting that the simultaneous coexistence of all perceptory states within perceptual infinity is structurally complete and unified. See Appendix B.

Connection / disconnection. A paired field-formation dynamic involving the making and breaking of entanglement between Perceptory Fields. See Chapter 3.

Conservation / objectification. A paired field-formation dynamic by which Perceptory Fields stabilize their state against perturbation (conservation) or reify their pattern into persistent artifacts (objectification). See Chapter 3.

Consensus reality. The shared registration-pattern that emerges when multiple Perceptory Fields entangle their internal models within a Perceptory Network, with each field registering the others' models. Not metaphor: structurally distinct from any individual field's private perception. See Chapter 3; Chapter 4.

Containment. The asymmetric form of entanglement: one Perceptory Field held within another, so that an inside, an outside, and a level appear. Containment is what builds hierarchy (cell within tissue, atom within molecule, proton within atom, quark within proton), as distinct from entanglement proper, which fuses peers symmetrically into a single unit. See Law VI.

Consciousness. In the framework's structural sense: the closed feedback loop that constitutes any Perceptory Field's self-perception. Operates at every scale of the hierarchical entanglement chain wherever stable closure obtains. Whether and to what degree this structural feature corresponds to subjective experience as humans encounter it varies by substrate. See Chapter 4.

Contraction / expansion. A paired field-formation dynamic by which Perceptory Fields concentrate or distribute their perceptory loops. See Chapter 3.

Cosmic web. The filamentary geometry of galaxy distribution at cosmological scales. The framework predicts that the cosmic web exhibits structural regularities (three-fold junctional preference, hierarchical self-similarity) consistent with the Rule of Three operating at cosmological substrate. See Law IX; Chapter 2; Appendix C, Prediction 4.

Disintegration. The failure of three-point closure in a Perceptory Field, with components separating into smaller Perceptory Fields that retain structural integrity at lower levels. Distinguished from dissolution. See Chapter 3.

Dissociation / reorganisation. A paired field-formation dynamic by which a Perceptory Field splits into parallel sub-loops (dissociation) and re-merges in new configurations (reorganisation). Trauma is read structurally as dissociation whose reorganisation has not occurred. See Chapter 3.

Dissolution. The return of a Perceptory Field to the symmetric ground state, the Perceptory Field of Nothingness, with no smaller fields preserved. Distinguished from disintegration. Most observable transformations are disintegrations, not dissolutions. See Chapter 3.

Entanglement. The condition under which two or more Perceptory Fields are in mutual registration through Boundary breach; a change in any entangled field produces a change in the others through the breach. In its symmetric form it fuses peers into a single unit; its asymmetric form is containment. See Law V; Law VI.

Entangling / disentangling. The deeper-level antipodal-pair structure of gravity: not gravity-versus-anti-gravity, not attraction-versus-repulsion, but the entangling and disentangling phases of network interaction held in inseparable relation by the perceptory boundary between entangled and disentangled states. See Law VIII; Chapter 1.

Expression / camouflage. A paired field-formation dynamic by which Perceptory Fields broadcast or suppress their state to surrounding fields. See Chapter 3.

Hierarchical Entanglement. The architecture by which Perceptory Fields organize through two relations: entanglement (symmetric, peers fused into a single unit) and containment (entanglement gone asymmetric, one field held within another, producing levels). Hierarchy is containment stacked upon containment, the directed form of entanglement rather than a principle imposed from outside. See Appendix A; Law VI.

I and Me. Mead's distinction, adopted by the framework as the cognitive-scale instance of the structural Perceptory Boundary. The I is the subjective perceiver; the Me is the perceived self; the Boundary between them is the dynamic feedback loop that makes self-awareness possible. See Law IV; Chapter 2; Chapter 4.

Infinity identity (I(id) = 1). A structural postulate (Appendix B) asserting that the totality of perceptual infinity, in its self-identical aspect, is structurally singular and unified. See Appendix B.

Inseparation Constant. The structural rule that governs the boundary of every Perceptory Field that has departed from the perfectly symmetric ground state. Wherever a Perceptory Field has an asymmetric boundary, the boundary produces an inseparable pair, perceived and not-perceived, that come into existence together and cannot be separated without dissolving the field that contains them. See Appendix A; Law IV.

Inseparation Element. The structural object that holds an inseparable pair in their inseparable relation. The third element of three-point closure. Specific instances: the A-particle

(particle/antiparticle); the magnetic field (north pole/south pole); the electromagnetic field (positive/negative charge); the bond (mother/child); the act of perception (self/other); the Perceptory Boundary (I/Me); the perceptory boundary between entangled and disentangled states (gravity); the Perceptory Force (perceptual infinity/propagation infinity). See Appendix A; Law VI; Chapter 2.

Inseparation Element of gravity. The perceptory boundary between entangled and disentangled states; the threshold at which a Perceptory Network either holds its entanglements or releases them. See Law VIII; Chapter 2.

Networks. See Perceptory Network.

Nothingness. The absence of anything; non-dimensional, non-interacting, infinite in its incapacity, structurally prior to all existence. The unperceiving ground. Not Infinity. See Appendix A; Law I.

Parity restoration. The framework's predicted condition at high energies under which the suppressed right-handed weak interaction is amplified to equality with the left-handed weak interaction, restoring parity that is broken at low energy. See Law VII; Appendix C, Prediction 2.

PF. Abbreviation for Perceptory Force.

PF-mode. A specific manifestation of the Perceptory Force operating under specific dimensional conditions. The framework recognizes magnetism as the 1-D PF-mode, electromagnetism as the 2-D PF-mode, the strong nuclear force as the manifestation of three-point closure at the matter scale, and the weak nuclear force as antipodal-pair structure with amplitude asymmetry. Gravity is treated as a different kind of force operating between Perceptory Networks rather than within a single field. See Chapter 1.

Perceptory Boundary. The structural element of a Perceptory Field that distinguishes its interior from its exterior. Asymmetric, located at the closure-perimeter, dynamic rather than static; an active feedback loop that detects, compares, and adjusts. The structural location at which the Inseparation Constant operates. Breachable. See Appendix A; Law IV.

Perceptory Field. The structural object produced when the Perceptory Force closes a self-perceiving loop. The category applies at every scale of the framework. See Appendix A; Law III.

Perceptory Field of Nothingness. The first instance and the ground state of the Perceptory Field category. Infinitely symmetric, perfectly stable, prevalent throughout infinity, and dispersed through infinity. Closed and not open: it has no exterior, because it encapsulates in every direction the Nothingness any exterior would be made of, and it is one, not many. See Appendix A; Law III.

Perceptory Force (PF). The fundamental act through which existence is registered. Structurally prior to the four forces of mainstream physics. The Inseparation Element of perceptual infinity and propagation infinity. See Appendix A; Law II.

Perceptory Network. Two or more Perceptory Fields in entangled relation through mutual breaching of their Perceptory Boundaries. The substrate of all observable interaction. See Appendix A.

Perceptory Sphere. The structural representation of perceptual infinity: a totality carrying antipodal pairing (every distribution paired with its opposite) without any metric, topology, or spatial extent. The earlier zero-volume, zero-radius notation is withdrawn in Appendix B; the sphere borrows only the antipodal-involution structure, not Euclidean geometry. See Law X; Appendix B.

Perceptual Infinity. The totality of all perception observed as totality. Symmetric, closed, dimensionless. The same structural object as the Perceptory Field of Nothingness considered at the largest scale. See Appendix A; Law X.

Propagation Infinity. The totality of all asymmetric extension. Open, ongoing, never resolving into stable equilibrium. Time is what propagation infinity feels like to a Perceptory Network from inside. See Appendix A; Law X.

Replication / propagation. A paired field-formation dynamic by which Perceptory Fields create copies of themselves (replication) or spread their pattern through adjacent fields without copying (propagation). See Chapter 3.

Rule of Three. The geometric condition under which Hierarchical Entanglement produces stable Perceptory Fields. Three is the minimum number of entanglement points that produces a stable closed Perceptory Field, with the three points configured as two inseparable opposites held by an Inseparation Element at the centre. See Appendix A; Law VI; Chapter 2.

Rule of Three identity (H(id) = 3). A structural identity, derived as a theorem in Appendix B, stating that the minimum stable closure is three-point structure, with three as the structural identity of stable closure at every substrate. (Labelled the Hierarchical Entanglement identity in earlier drafts.) See Appendix B.

Substrate-relativity. The framework's claim that time, gravity, and consciousness are all relative to the substrate of the Perceptory Network at which they are perceived or operate. Each substrate registers change, gravitates, or perceives at its own rate, with no privileged or universal timescale. See Law VIII; Law X; Appendix C, Prediction 9.

Three identities, the. The framework's three structural constants: Infinity identity (I(id) = 1), Coexistence Constant (C(id) = 1), Rule of Three identity (H(id) = 3). See Appendix B.

Time. What propagation infinity feels like to a Perceptory Network from inside; substrate-relative; the perception of asymmetric change registered by the network's local feedback loops. See Law X; Chapter 3 (acceleration/deceleration); Appendix C, Prediction 9.

Wave-function collapse. The reconfiguration of an Inseparation Element when a fourth perceptory entanglement (the observer's Perceptory Field) is introduced into a three-point closed structure (the wave-particle superposition). Unlike pair-annihilation, where the pair itself dissolves, the pair members persist here and the

Element re-forms within the enlarged Field that now includes the observer; reconfiguration and dissolution are the two outcomes of fourth-entanglement breach, distinguished by whether the original pair survives it. See Law VI; Chapter 2.

Appendix E, Relation to Existing Theories

The framework of the Laws of Consciousness does not propose itself as a replacement for existing theoretical work. It proposes itself as a structural account that re-frames what existing theories have documented and connects them within a single architecture. This appendix locates the framework relative to the principal theoretical traditions it engages with, in consciousness studies, in fundamental physics, in philosophy of mind, in cosmology, and identifies, for each, the specific points of convergence, divergence, and extension.

The appendix is offered for two purposes. First, to allow readers familiar with each tradition to assess the framework's position rapidly, with explicit signposts to where the framework agrees, disagrees, and extends. Second, to invite engagement from practitioners within each tradition, with the understanding that the framework's structural categories may either illuminate or be illuminated by the work each tradition has already done.

Integrated Information Theory (IIT)

Giulio Tononi's Integrated Information Theory proposes that consciousness corresponds to integrated information (Φ) within a system, with consciousness identified with the quantitative measure of how integrated the system's information processing is. IIT has been developed mathematically in considerable detail and has generated specific empirical predictions about which neural systems should and should not produce consciousness.

Convergences. The framework converges with IIT on the structural identification of consciousness with closed integrated systems, feedback loops that integrate information across multiple components. The framework's reading of consciousness as the closure of feedback loops within Perceptory Fields (Chapter 4) is structurally adjacent to IIT's identification of consciousness with integrated information. Both treat consciousness as a structural feature of certain kinds of system architecture rather than as an epiphenomenon or as a mysterious additional substance. Both are willing to attribute consciousness, in their respective senses, to systems beyond brains.

Divergences. The framework diverges from IIT on the claim that integrated information is itself the substrate of consciousness. The framework reads integrated information as the visible signature of the structural feedback that closes a Perceptory Field, not as the substance of consciousness itself. Consciousness, in the framework's structural sense, is what closure looks like from inside; integrated information is what closure looks like from outside, when measured externally. The two are different perspectives on the same structural feature.

Extensions. The framework extends IIT in several directions IIT has not pursued. The framework derives consciousness from the structural primitives of perception, Perceptory Field, and Perceptory Boundary, providing a deeper architecture from which integrated information can be understood as one signature among others. The framework extends consciousness to substrates IIT has not addressed (gravitational, cosmological, the level of the totality). And the framework offers specific predictions, the A-particle, parity restoration, cosmic-web structural regularities, that IIT does not generate.

Global Workspace Theory

Bernard Baars' Global Workspace Theory proposes that consciousness arises when information is broadcast across a global workspace within the brain, making information available to multiple specialized processors simultaneously. Stanislas Dehaene's neuronal version (Global Neuronal Workspace) has developed the theory in neurobiological terms, identifying consciousness with the ignition of widely distributed cortical networks.

Convergences. The framework converges with Global Workspace Theory on the importance of integration across distributed systems. The framework's reading of consciousness as the closure of feedback loops within Perceptory Fields, and of larger-scale consciousness as the closure of feedback loops across Perceptory Networks (Chapter 4), captures structurally what Global Workspace Theory captures functionally. Both treat consciousness as requiring widespread integration rather than localized processing.

Divergences. The framework diverges from Global Workspace Theory on the locus of consciousness. Global Workspace Theory localizes consciousness to the workspace itself, the cortical broadcast architecture. The framework reads consciousness as a structural feature present at every scale where stable closure obtains, not as a property of one specific architectural feature within brains. The framework's account of molecular, cellular, and organismal consciousness has no counterpart in Global Workspace Theory.

Extensions. The framework extends Global Workspace Theory by providing the structural reason why workspace-style architecture produces what it does. Workspace-style architectures produce consciousness, in the framework's reading, because they instantiate the Perceptory Boundary's feedback closure at the cortical scale. Other architectures, at other substrates, produce other forms of consciousness, the framework's structural sense, through different physical instantiations of the same structural feature.

Predictive Processing and the Free-Energy Principle

Karl Friston's free-energy principle, developed extensively over the past two decades, proposes that biological systems minimize prediction error across all their sensory and motor modalities, with consciousness arising from the dynamics of prediction-error minimization in sufficiently complex systems. The closely related predictive processing framework (Andy Clark, Jakob Hohwy, and others) has developed similar architectures in cognitive-science terms.

Convergences. The framework converges with predictive processing more closely than with any other consciousness theory currently in the field. The framework's structural definition of consciousness as the closed feedback loop that detects, compares, and adjusts (Law IV; Chapter 4) is structurally identical to the predictive-processing architecture of generating predictions, comparing them with sensory input, and adjusting model and behaviour to reduce discrepancy. The framework's reading of Friston's free-energy principle as the structural feedback loop of Law IV operating at the neural substrate (Chapter 4) is explicit and direct.

Divergences. The framework diverges from predictive processing on the question of what is being predicted and why. Predictive processing typically frames prediction-error minimization as a biological-functional principle: organisms minimize prediction error because doing so contributes to evolutionary fitness. The framework reads the same architecture as the structural condition for consciousness itself: the closure of the prediction-comparison-adjustment loop is what consciousness structurally is, with biological-functional advantages following from rather than grounding the structural feature.

Extensions. The framework extends predictive processing across substrates predictive processing has not addressed. Molecular consciousness (protein folding as conformational prediction-error minimization), cellular consciousness (homeostatic regulation as cellular-scale prediction-error minimization), and social consciousness (Perceptory Networks producing consensus reality through distributed prediction-error minimization) are all framework instances of the same architecture. Predictive processing has been developed primarily for neural and cognitive systems; the framework predicts that the same structure will be visible at every substrate where stable Perceptory Fields exist.

Panpsychism

Panpsychism, in its various forms (constitutive, cosmopsychism, Russellian monism, and others), proposes that consciousness is a fundamental feature of the universe, distributed in some form across all matter or all reality. The contemporary revival of panpsychism (Galen Strawson, Philip Goff, David Chalmers in his recent work) has developed sophisticated philosophical arguments for the view.

Convergences. The framework converges with panpsychism on the rejection of consciousness as exclusively localized to brains. Both treat consciousness as a feature distributed across a wider domain than mainstream materialism allows. Both face the question of how the consciousness of small things (atoms, particles, simple systems) relates to the consciousness of larger things (organisms, minds, societies).

Divergences. The framework diverges from panpsychism on the structural ground of consciousness. Panpsychism typically treats consciousness as a primitive feature of fundamental constituents, atoms, fields, or basic particles, and asks how these primitive consciousnesses combine into more complex consciousness. The framework treats consciousness as a structural feature of any closed Perceptory Field, with the closure being primitive rather than the consciousness. The framework does not require that fundamental particles have consciousness in the panpsychist sense; it requires that wherever a stable Perceptory Field exists, the structural conditions for consciousness obtain.

Extensions. The framework extends panpsychism by providing the structural mechanism panpsychism typically lacks. Panpsychism's combination problem, how do small consciousnesses combine into large ones, is addressed structurally by the framework: containment (Law VI), entanglement gone asymmetric, produces nested Perceptory Fields whose consciousnesses are themselves nested, with the larger field's consciousness emerging from the closure of feedback across its constituent fields. The framework also identifies specific predictions where panpsychism makes none, and offers itself for falsification at those points.

Neutral Monism

Neutral monism (William James, Bertrand Russell at certain points, and a contemporary lineage including David Chalmers' double-aspect monism) proposes that the fundamental ground of reality is neither material nor mental but a third kind of stuff that grounds both. Mind and matter, on this view, are two aspects of the same underlying neutral reality.

Convergences. The framework converges with neutral monism on the rejection of the material-mental dichotomy as foundational. The framework's structural account of perception, Perceptory Field, and consciousness does not begin from material entities and add mental properties, nor does it begin from mental entities and add material properties. It begins from perception itself, with material and mental phenomena both read as substrate-specific manifestations of the same structural primitives.

Divergences. The framework diverges from neutral monism on the character of the neutral ground. Neutral monism typically posits a third kind of stuff, neither material nor mental, that grounds both. The framework posits perception itself as the structural primitive, with the Perceptory Force as the fundamental act through which existence is registered. This is not a third kind of stuff; it is the act through which any kind of stuff comes to exist as registered. The framework's primitive is structural and functional rather than substantial.

Extensions. The framework extends neutral monism by providing the structural architecture that neutral monism typically lacks. Where neutral monism asserts that material and mental phenomena are aspects of a neutral ground, the framework specifies how: through the Perceptory Force operating at different substrates, with substrate-specific Perceptory Fields manifesting as either material or mental depending on the scale and substrate of observation. The framework's specific predictions and falsification conditions also go beyond what neutral monism typically commits to.

The Participatory Universe

John Wheeler's participatory universe principle proposes that the universe requires observers, that the act of observation is constitutive of physical reality rather than merely revealing pre-existing reality. Wheeler's reading of quantum mechanics, particularly the delayed-choice experiments and the it-from-bit programme, treats observation as fundamental rather than incidental.

Convergences. The framework converges with Wheeler's participatory universe on the foundational claim that perception is constitutive of existence. Law I's keystone derivation, that for Nothingness to exist, it must be perceived, is structurally consistent with Wheeler's it-from-bit. Both treat observation/perception not as an external relation to a pre-existing reality but as the act through which reality comes to exist as registered.

Divergences. The framework diverges from the participatory universe on the scope and substrate of perception. Wheeler's framework is typically read as referring to observation in the sense of measurement, interactions involving measuring apparatus or, ultimately, observers in the conventional sense. The framework reads perception more broadly as the structural act that occurs wherever any Perceptory Field closes a feedback loop, at any substrate. Perception in the framework's sense does not require an observer in Wheeler's sense; it requires only the closure of a self-perceiving loop.

Extensions. The framework extends Wheeler's programme by specifying what perception structurally is and how it operates across substrates. Where Wheeler offered the participatory principle as a foundational commitment, the framework develops the structural architecture, perception, Perceptory Field, Perceptory Boundary, Inseparation Constant, Inseparation Element, Hierarchical Entanglement, antipodal-pair structure, that the principle implies. The framework's predictions follow from this architecture; Wheeler's programme generates few specific predictions of its own.

Entropic Gravity and Emergent Spacetime

Erik Verlinde's entropic gravity programme, Mark Van Raamsdonk's work on the AdS/CFT correspondence and entanglement-based spacetime emergence, Ted Jacobson's earlier work deriving Einstein's equations from thermodynamic principles, and related contemporary work all propose that gravity and spacetime are not fundamental but emerge from underlying entanglement or informational structure.

Convergences. The framework converges with entropic-gravity programmes on the architectural claim that gravity emerges from entanglement structure. The framework's reading of gravity as the act of perceptory entanglement between Perceptory Networks (Law VIII) is structurally adjacent to Verlinde's entropic-gravity proposal. Both treat gravity as derivative rather than primitive. Both read spacetime geometry as a consequence of underlying entanglement architecture rather than as a fixed background.

Divergences. The framework diverges from entropic-gravity programmes on the substrate-extending claim. Verlinde, Van Raamsdonk, Jacobson, and others develop emergent gravity within the matter-scale substrate, treating quantum entanglement among matter-scale degrees of freedom as the foundation from which spacetime and gravity emerge. The framework reads the same architectural insight as substrate-general: gravity emerges from entanglement at every substrate of the perceptory hierarchy, with matter-scale gravity being one specific case rather than the foundational case.

Extensions. The framework extends entropic-gravity programmes by predicting that accumulative attraction, the opposite-less drawing-together Law VIII distinguished from the antipodal attraction of Law VII, operates at substrates those programmes have not addressed. Cognitive, biological, and cultural accumulation, each is, in the framework's reading, the same emergence-from-entanglement mechanism operating beyond the matter scale (Law VIII; Chapter 1; Chapter 5), while the antipodal attractions at those substrates belong to Law VII and are not gravity. The framework also offers specific predictions about black hole dynamics, the in-fall of matter and the bipolar jets it launches as the entangling and disentangling phases, and about the antipodal-pair structure of gravity that emergent-gravity programmes have not yet developed.

Eastern philosophy and contemplative traditions

The framework has structural affinities with several long traditions in Eastern philosophy and contemplative practice, affinities the appendix names without attempting to resolve. Buddhist philosophy's account of dependent origination (pratītyasamutpāda) has structural parallels with the framework's treatment of perception as constitutive of existence and of the inseparability of perceived and not-perceived (Law IV). Vedantic non-dualism has affinities with the framework's reading of perceptual infinity as structurally singular (Appendix B's I (id) = 1). Daoist accounts of the dao as the ungrounded ground from which manifest reality unfolds have affinities with the framework's reading of Nothingness and the perception of Nothingness.

The framework does not claim to derive from these traditions, nor does it claim to translate them into structural-physical terms. It notes the structural affinities and acknowledges that the traditions have developed sophisticated phenomenological and philosophical analyses of structurally similar territory over millennia of disciplined practice. Engagement between the framework and these traditions is offered as an open invitation rather than as a settled conclusion.

What this appendix has established

The framework of the Laws of Consciousness positions itself among existing theoretical traditions as a structural account that converges with each tradition at specific points, diverges at specific points, and extends each tradition in ways the tradition itself has not pursued. The framework converges most closely with predictive processing and entropic-gravity programmes, less closely but substantively with IIT, Global Workspace Theory, and panpsychism, and at the level of foundational orientation with Wheeler's participatory universe and certain Eastern contemplative traditions.

The framework offers itself for engagement with practitioners of each tradition. Where convergences are real, the framework predicts that detailed work within each tradition will continue to reveal structural features the framework also predicts. Where divergences are real, the framework offers itself for falsification at the specific points of difference, with the understanding that any tradition's empirical or theoretical successes that contradict the framework's predictions would constitute evidence against the framework rather than evidence for it.

The framework is not a synthesis of existing theories. It is a structural account that re-frames what existing theories have documented, names what they have left implicit, and connects what they have addressed separately. Whether the framework's structural account survives the engagement is a question the engagement itself must settle.

www.ingramcontent.com/pod-product-compliance
Lightning Source LLC
LaVergne TN
LVHW050911210926
848482LV00012B/398